The Unfinished Quest

The Plight of Progressive Science Education in the Age of Standards

The Unfinished Quest

The Plight of Progressive Science Education in the Age of Standards

by

Clair T. Berube
Hampton University

Information Age Publishing, Inc.
Charlotte, North Carolina • www.infoagepub.com

Library of Congress Cataloging-in-Publication Data

Berube, Clair T.
The unfinished quest the plight of progressive science : education in the age of standards / by Clair T. Berube.
p. cm.
Includes bibliographical references.
ISBN 978-1-59311-928-7 (pbk.) -- ISBN 978-1-59311-929-4 (hardcover) 1. Science--Study and teaching--Standards--United States. I. Title.
Q183.3.A1B475 2008
507.1'073--dc22

2008017756

ISBN 13: 978-1-59311-928-7 (pbk.)
978-1-59311-929-4 (hardcover

Printed in the United States of America

For Mo with love, husband, colleague, friend

CONTENTS

ACKNOWLEDGMENTS

I would like to thank several people who have helped me along the journey that led up to this book. First and foremost I would like to thank my husband, Dr. Maurice R. Berube, for his support over the years, both intellectually and personally, and for his fine editing, proofreading, and unsurpassed intellectual insight. It is a wonderful thing to be married to someone who appreciates and understands what you do. I would also like to acknowledge the following, for their support when the going got tough, they kept me going; Lee Manning, Dwight Allen, Steve Tonelson, Bill Cunningham, SueAnne McKinney, all of Old Dominion University in Norfolk, Virginia, Jeffrey Glanz of Yeshiva University in New York who believed in me enough to give me my first real academic job, Dr. Rebecca Bowers, my first mentor and dean of education at Central Washington University, and to all of my other friends, academic and otherwise, who gave words of support and encouragement along the way. This book has traces of them in it.

INTRODUCTION

I taught middle school science for 5 years. During this time I met many students, some of whom knew how to think and some who did not. Many had been trained by modern society to seek what was expected and to deliver exactly that, nothing more, nothing less. Many only wanted to know what I wanted, so that they could make me "happy" and therefore acquire a good grade. Indeed, grades were what made their worlds go 'round. Sometimes, I would assess their understanding with no grades attached. Most of them did not like this at all, and wanted to "get something" for their troubles, and that something was a grade. But there were also some students who argued with me, came up with new ways of solving problems, who were actually delighted when they were given something to figure out, and who displayed their teenage rebelliousness through their intellect, which is the very best kind of rebellion indeed. My goal as a science educator became then, and still is now, to uncover each science student's sense of rebelliousness and authority questioning (critical thinking skills) hidden deep within them. The best scientific discoveries were achieved through this mindset, instead of the current educational sheep herding mentality, which is rewarded and is so prevalent in the United States today. Our schools reflect the society we live in, and we are being led around by our necks like sheep in the dark.

As a result, American science education has lost its way. Modern educators are currently ideologically in one of two camps: those who see American education in general as heading in the right direction, and the ones who fear that it has gone tragically astray. For over 100 years the American educational system has been the hope of trust-fund baby and immigrant alike; the most important means we have as a Nation to level the playing field of opportunity. Yet we continue to lag behind several other industrialized countries when comparing standardized test scores. Is this acceptable in the most advanced, affluent country on earth? Why are we not the reigning educational system on the planet? And are standardized test scores even the best way to assess the type of learning it takes to lead the world in science discovery and innovation?

The twentieth century started out well enough, with high ideals and firm direction, but as the latter part of the century unfolded, we found ourselves bogged down in the very "standards movement" that sought to raise us up as the leaders in the world. While needing high standards, we somehow became misguided in our attempts to enforce those standards, to the point of losing any real educational reform (reform aimed in the positive direction anyway). The nature of science does not lend itself well to bubble tests, but these assessments are all we have to "prove" that our students are good scientists.

One wake-up call came on the morning of September 11, 2001, when airplanes were used as bombs in the single most horrific act of terrorism ever seen. That we did not see it coming, and that we were amazed at the terrible brilliance of it, are signals that we Americans can be less creative than our enemies, and that was a shock indeed, unprecedented in American history. What if changing the way we teach and assess learning better prepares our citizens for creative preventions and solutions to world problems, instead of creating citizens who have to react to world problems? Do we already have creative brilliant minds in America? Of course we do. Do we already have good schools that employ progressive teaching and learning pedagogy at the highest levels of learning? Yes to that also, but many of these schools are limited to the wealthiest suburbs or to private prep schools.

This book will address these issues, as I will also examine those who led the way for true educational reform, those responsible for forcing it off-track, and what can be done to correct the problem. Can we as a nation truly educate all of our citizenry, while at the same time holding our teachers and students to high-quality educational outcomes? We can, but a clear understanding of what we are seeking at our educational finish-line is important. Unfortunately we have lost site of that too.

CHAPTER 1

THE PROGRESSIVE MOVEMENT AND SCIENCE EDUCATION

"In tomorrows' enterprise the knowledge worker will be freed to release creative energy that will result in an era of enormous innovation and discovery, fulfilling the potential and promise of the mind."

—Frederick Winslow Taylor (1911)

The term "progressive" refers to many things, but in education, it is used to denote a philosophy that emphasizes the individual child, and encouraging self-expression (Miriam-Webster, n.d.). Today, American education is mired at best, and going backwards at worst. The reason? Since the dawn of the "Standards Movement" in the early 1980s (which itself was driven by the U.S. government publication of "A Nation at Risk" [1983]), the American educational system has reverted to high-stakes, multiple-choice tests to raise the so-called "standards" which have been set forth for each state. In doing so, we as a nation have used the evidence of multiple-choice test scores as evidence of success; supposedly meaning better teaching and learning are taking place in our schools. Is this the case? No one is arguing that higher standards are not in order, but are our measuring tactics for these standards accurate? Reliable? Ethical? What do

The Unfinished Quest: The Plight of Progressive Science Education in the Age of Standards
pp. 1–7

higher multiple-choice test scores really tell us and what does it really mean?

The American educational system is a fairly recent institution. And if we think of the changes our system has undergone since its inception, it is remarkable. Until the second half of the nineteenth century, Americans were rural people. Towards the end of the century, the population of urban dwellers more than doubled, approaching 50% (Urban & Wagoner, 1996, p. 59). As farmers migrated to large cities, the class structure became more evident, resulting in a new lower and upper class system.

> In 1890, 11 million of the nation's 12 million families earned less than $1200 per year; of this group, the average annual income was $380, well below the poverty line. Rural Americans and new immigrants crowded into urban areas. Tenements spread across city landscapes, teeming with crime and filth. Americans had sewing machines, phonographs, skyscrapers, and even electric lights, yet most people labored in the shadow of poverty. (Hoyt, 1999)

With the advent of the Gilded Age, came a host of societal adjustments. An economist in 1879 noted "a widespread feeling of unrest and brooding revolution." Violent strikes and riots wracked the nation through the turn of the century. The middle class whispered fearfully of "carnivals of revenge" (Hoyt, 1999).

After the Civil War, industry was booming, thanks in part to reconstruction of the South. Mass production was also producing mass wealth for the few who became the titans of industry. With mass migration to the cities to man these factories, the distance between the classes became even more evident. While the bosses' wealth grew, the workers fell towards poverty, giving rise to unrest and the beginning of trade unions. During this time, an inventor, Frederick Winslow Taylor, developed a management system and a theory of organizations that ultimately had an impact on American schools as well as industry. His theory included clear delineation of authority, and task specialization. He also developed the "efficiency movement," which was concerned with making business more cost-efficient. Schools were looked upon as workplaces and were measured in terms of productivity (Barger, 2004). Elwood Cubberly, a turn-of the century historian, stated that schools should be like factories. Believing that the mission of the public schools were to: "assimilate the new immigrants into a nation that would remain English speaking and thinking," he was quoted as saying that public schooling would implant "The Anglo Saxon conception of righteousness, law and order, and popular government' into the immigrant children" (Barger, 2004). The children who could not be processed to completion were considered as scraps. Therefore they were considered to be dropped out of the production line which is where we get our most accurate definition of "drop outs" (Barger, 2004). These things in turn, affected the schools, which were beginning to reorganize. School systems were formed, where schools were parts of larger run organizations with central office administrations.

Also a result of reconstruction, each state was to provide a system of free public schooling for its citizens. Some states like Texas, refused at first, claiming that if they were to provide free schools, they would be segregated. If such states sent their state constitutions to Congress for approval, Congress would not allow those southern states to rejoin the union until they had agreed to provide free unsegregated public schooling, sending their constitutions back to the state for revision. As a result, 10 southern states had governmentally imposed free education, whereas before the war, the only free schooling in the South was for pauper students who couldn't afford an education (Urban & Wagoner, 1996, p. 161).

During post-Civil War America, the executive branch of the federal government created a federal bureau of education, and even though the role of this bureau was mostly data gathering, it brought with it the most governmental intervention and involvement than had been experienced so far and heralded the dawn of governmental control over American education. The first arena in which this was to become evident was compulsory school attendance laws, enacted in twenty-seven states by 1890. By 1918, all 48 states were to have such legislation. This was a giant step forward towards protecting children from the inhuman working conditions many of them were forced to suffer in postindustrial America, even though the idea of education at this point, was still to provide skilled workers. There was some opposition to this law, coming from the more rural states where child labor was often the difference between a family eating or starving (Urban & Wagoner, 1996, p. 163).

With growing political awareness, American schools also began to be charged with creating good citizens, as well as workers. Indeed, the notion that democracy would not survive without an educated populace began to shape curricula for the schools. The voting public had to have a good education in order to send the politicians and leaders of that democracy into office. Without knowledge of politics, civics, and American history, the best candidates would not be elected, thus holding back American progress towards the new century.

Stabilizing society through education was another goal of the American system during this time. The so called "stable" American society was under threat by many real and imagined forces, including immigration, ethnic conflict, crime, and class divisions. The school with its emphasis on social conformity and discipline, was thought to be the panacea of these societal ills (Urban & Wagoner, 1996, p. 169). Social Justice theories came into existence, with the main goal being to provide a venue (education) whereby immigrants and those of similar cultural status, could through education, rise in the ranks of society to achieve the American dream.

JOHN DEWEY

One of the most important persons in American education was John Dewey, philosopher and educator, and one of the first proponents of social justice theory as it pertains to classroom education. Much has been written about Dewey, and he will be discussed again in the next chapter. But one cannot consider progressive science education or progressive education at all without John Dewey. His views revolutionized education in the United States, so much so that today in college classrooms across America, he is often called the father of American education.

Dewey was a Yankee, born in Vermont in 1859, the son of a society marm and a farmer. He was a good student, but abhorred the memorization that was mandatory in much of school. It was in college studying philosophy that he realized that so much of a person's education happens outside the classroom, and this view played a large role in his later educational philosophies, including constructivism. The list of agencies and institutions he touched and changed are long and distinguished. He was a founding father of the American Federation of Teachers while still a member of the National Education Association. He was instrumental in founding the New School for Social Research and led the way for the creation of several lab schools around the country; schools that experimented with various progressive teaching pedagogies and philosophies (M. Berube, 2000).

The reason Dewey is so important in the progressive movement, is that he truly had revolutionary ideas concerning children and learning. He reasoned that children are not one-dimensional, but should be educated in four areas instead of one: intellectual, moral, social, and artistic. This was quite a disconcerting concept to most educators in the age of "children should be seen and not heard." This "whole child" concept threatened not only to overthrow conventional educational practices, but how children should be raised and treated as human beings as well. This opened up the idea of two important aspects of learning, not only nature, but nurture. Also, one could not now study education without having thorough knowledge of human development and psychology and their relationship to learning.

Another groundbreaking idea of Dewey's was the notion of learning being rooted in experience. Experiential learning had not up to that point been viewed as part of the necessary classroom canon, since much of the purpose of education at the time was to *remove* the students from their daily duties on the farm, sweatshops, or jobs. Whatever the children learned via these experiences was not deemed important or culturally significant, and could not tie into classroom experience. Dewey saw the process of thinking as problem solving, and that all learning should teach the child to be an independent thinker who is able to solve problems and to think critically. Indeed, Dewey saw an organic connection between education and personal experience. But the quality of the experience was important, stressing process over product, with the idea that the product would be sound if the process was sound. This was a very important realization, since many times a product appears to be fine, but in reality is flimsy or unsound due to faulty process in its creation. This notion was no doubt born from Dewey's Yankee pragmatic lifestyle, but one well suited to developing intelligent, independent children (M. Berube, 2000, p. 40). Indeed, Dewey epitomized the American idea of education, including the idea that the purpose of American education was to democratize all citizens, and to produce able, intelligent progressive problem-solving members of the population, who could make well-informed decisions about their lives and country. Education was to be a freeing experience, emphasizing freedom of thought on the part of the learner without the external imposition traditional schooling offered that limited intellectual and moral development of the young.

Since Dewey, education has seen many relevant philosophers and leaders who seem to perpetuate and add to the Dewey educational philosophy. Many of the components of successful classrooms today came from Deweyan ideas, such as cooperative learning, moral and arts education, and problems-based education. American education would not be recognizable without the enormous impact of John Dewey.

In the science classroom, Dewey has had a huge influence also, though I doubt he could have foreseen this since he was not a scientist but a philosopher. Many of the components of education that Dewey lauded are evident in a well-run science classroom, including discovery learning, experiential learning including labs and hands-on activities, cooperative learning (which incorporates Dewey's social learning theory), and creative, inductive learning which is vital for scientific research. More will be said about this later, but the very foundations for successful science learning started with John Dewey.

There were other important social reformers on the horizon during this time that affected American society, and in doing so, also affected American schools. One was Jacob Riis, a former police reporter for *The New York Tribune* who became sympathetic towards the poor during his journalism career. His famous book *How the Other Half Lives* was published in 1890 and described his romantic view of poverty while calling the poor "Noble savages" and "the victims of society." According to Riis, poverty did not involve a dark side to human character, but only a victimization of those involved, due to lack of economic opportunities and discrimination. This view contradicted the traditional conservative notion of the poor as lazy, stupid, or lacking in character in some way. Instead the Romantics viewed the poor as having inherent goodness. Riis spoke of the poor having "no way out" because of the system, and that children were the key to social reform and to curing poverty. If housing problems could be solved he argued, there would be no poverty. The Romantic view of social reformers impacted education in ways, including striving for equal education for all, regardless of income or class.

Another important social reformer who influenced American education at the turn of the twentieth century, was Jane Addams, who is most famous for founding Hull House, the American settlement house in the Chicago slums in 1889. Disturbed by how the poor were treated, she decided they needed a place to go; a center that provided for a higher civic and social life, just like middle class people. Hull House had a co-op for coal and sugar, started the country's first kindergarten, offered classes in sewing, music, and painting. It also had America's first public playgrounds, first coffee house (for an alternative to saloon life), and the first juvenile court. She believed firmly that poverty was not caused by vice and laziness, and that it was man's duty to help provide for those less fortunate. Although White, she founded NAACP (National Association for the Advancement of Colored People) and fought for women's suffrage. Her work lifted the spirits of the poor, and offered them hope and inspiration.

The third social reformer that directly impacted education was Walter Rauschenbusch, a theologian and New York City Baptist minister. Rauschenbusch taught the "Social Gospel" in his sermons, and proclaimed that one should direct his attention to the world and to the plight of poverty, rather than to heaven. He spoke of poverty being spiritually and physically numbing and that political equality was the basis for Christian morality (M. Berube, 1994, pp. 1-11).

The modern American school system still shows the effects of the courageous leaders who laid the foundations for the modern democratic educational system we value today, and its important place in society as the "great equalizer," Since schools are always a reflection of the society in which they reside, schools today can thank early social reformers for opening the way for a free public education for every American, regardless of immigrant status, wealth, class, race, or religious persuasion, which is the basis of a democratic society and a springboard towards the American Dream.

The question that naturally arises is this: can progressive science education and progressive education in general coexist with today's data-driven, standardized test based school system? Any student of education in any teacher's college across the country has been taught that you must "teach to the objective," in other words, plan your instruction so that by the end of the lesson, your students will be able to do what you stated in your objective. For example, if your daily learning objective is: "the student will be able to list and define the parts of the plant cell," then you teach them the parts of the plant cell, and you *assess* their knowledge of the parts of the plant cell. It would be unfair to assess anything other than what they have been taught. In this case, a simple multiple-choice (they do have their place in a good education!) or matching assignment would suffice.

On the other hand, if your daily objective is: "the student will be able to design a photosynthesis experiment demonstrating the components of photosynthesis," than the students obviously are beyond the stage of learning the parts of the plant cell, and are now working in higher levels of learning according to Bloom's (1956) taxonomy, including synthesis. As a teacher, you must assess this assignment differently than the former, and herein lies the problem with progressive education and standards. Although progressive education calls for higher-level learning and assessment, the standards movement, which supposedly calls for higher-level learning by holding teachers accountable for providing that learning, does not really stand for that at all, nor does it measure higher-level learning with the standardized test. And it is not the standards themselves, which are highly valuable. Those who support high-stakes assessments of the standards are more interested in assessing lower levels of learning to "prove" the students are learning and that "standards are being raised," than they are actually interested in the students learning anything. When the truth is told, higher-order thinking, problem solving and creativity are not important to powers that be. We have reduced our American educational system to the lowest common denominator, which we assess with the multiple-choice test.

Progressive education was born from the union of the desire to deliver social justice for all Americans through a quality education accessible to all citizens. Born of biblical teachings demanding those more fortunate among us to help those that are less fortunate, and refined in the ovens of slavery, Civil Rights, and the Holocaust, progressive education has a noble ancestry indeed but is now faced with a crisis; how best to educate American's children at the highest levels of thought and ability, while ensuring that all achieve the most basic standards? That which injures those with the least among us injures us all. The answers are apparent to any who will listen. This book will discuss what has happened to the progressive education movement as it pertains to science education since the age of standards and what consequences it holds for our future.

CHAPTER 2

CONSTRUCTIVISM AND SCIENCE EDUCATION

John Dewey never used the word "constructivism," nor did anyone else of his day. For at least a century before the term "constructivism" was coined, educators were implementing the philosophies and practices of the constructivist education movement under other names and philosophies. The leaders of the progressive movement spawned many such outgrowths, including social justice theory, constructivism, child-centered learning, and discovery learning. Today, constructivist learning philosophies drive not only many school districts' curricula and teacher practices, but many college and university teacher preparation programs as well, including some of the leading universities in the nation. It is fraught with controversy and disagreement among educators the world over, but it serves as a valid, highly effective model for educating the nation's children.

The idea that children build knowledge from their own experiences and mode of thought is the concept behind constructivism. It is the most important component of science discovery learning also. Another way of thinking about it would be that children and people in general, tend to extract different things from the same lesson or experience, depending on their world views and experiences prior to the learning. The coining of the term "constructivism" can be traced back to Piaget's reference to his views as "constructivist" and from Bruner's description of his discovery learning technique as "constructionist" (Applefield, Huber, & Moallen,

The Unfinished Quest: The Plight of Progressive Science Education in the Age of Standards
pp. 9–43

2000-2001). Those employing constructivist methodologies believe that real understanding occurs only when children participate fully in the development of their own knowledge, which occurs morally, cognitively, mentally, and socially.

> They describe the learning process as self-regulated transformation of old knowledge to new knowledge, a process that requires both action and reflection on the part of the learner ... the research of cognitive psychologists and science educators over the past decade has shown that what children learn greatly depends on what they already know. Knowledge and understanding grow slowly, with each new bit of information having to be fitted into what was already there. (Howe & Jones, 1993, pp. 8, 9)

Constructivism is also a philosophical explanation about the very nature of knowledge itself. As an epistemology, constructivism declares that knowledge is formed by the knower from existing beliefs and experiences. Knowledge is not independent of the knower and is not made up of accumulated "truths." Individuals create their own meaning from their own experiences; therefore, all knowledge must be tentative, personal, and subjective. Also, constructivism is an epistemological view of knowledge formation emphasizing construction rather than transmission and recording of information given by others (Applefield et al., 2000-2001).

Constructivism also can be defined as programs that are student-centered and are based on a theory of learning that focuses on how students develop understandings (Richardson, 1999). The constructivist approach differs from the traditional (direct instruction) approach in that students are included in the learning. Teachers who instruct from constructivist pedagogy develop lessons that lead children to engage in self-directed problem solving instead of direct instruction. Science classrooms which are taught by teachers who only lecture and "tell" are not conducive to student learning at the highest level. Science, more than any other subject, demands teaching styles that force the students to think for themselves, question conformity, and create their own learning.

> Most constructivists would agree that the transmission approach to teaching, usually delivered through lecture or direct instruction, promotes neither the interaction between prior and new knowledge nor the conversations that are necessary for intense involvement in ideas, connections between and among ideas, and the development of deep and broad understanding. (Richardson, 1999, p. 146)

Teachers assess the prior misconceptions that students bring to the classroom and try to correct them through this identification. Students use hands-on and cooperative learning situations and lessons that are student-centered based on children's basic curiosity about the world.

Also, constructivism is concerned with linking students' prior knowledge to present activities. According to McNichols (2000), "Constructivism is a theory about knowledge and learning." Embedded in this theory are the notions that:

1. Meaning, which is represented as knowledge, is based internally in the learner.
2. The acquisition of knowledge is the responsibility of the learner.
3. Knowledge is achieved from the learner's experiences and values conditioned by reflection, inquiry, and cognitive dissension.
4. Learning is an internal process, which is enhanced through the consensual negotiation of ideas.
5. The outcome of knowledge is a pragmatic process.
6. The assessment of learning is naturally connected with the learning process (McNichols, 2000).

These tenets of constructivism imply a classroom setting where social and intellectual interaction help students form meaning of the subject matter. In the science classroom, cooperative groups are the best way that students can tackle problems together to devise several solutions to one problem. Thus, constructivist pedagogy does not direct teachers in what and how to teach, but urges instructors to facilitate learning by providing a conducive environment for such in the classroom. Of course, many things go into successful teaching. There are old fashioned teachers who use lecture as the majority of their instruction, and who are very good at delivery and story telling, so that the students' attention is held. But again, in the science classroom, this is the least effective teaching technique. What constructivism seeks to add to the classroom experience, is the child in the role of his or her own educator, with the teacher as the guide. This does not take the teacher off the hook as the leader in the classroom. Nor does it ask of the children to learn the material themselves. Rather, instead of telling the students the answers that the teacher already knows, the teacher provides problems for the children to solve, thereby stressing process over product, and solidifying the learning in the children's minds. Things are better remembered when they are discovered and "worked out," rather than being passively received.

One of the biggest criticisms of constructivist pedagogy is that teachers are thought to have less content knowledge than teachers who use a more direct instruction approach. They are thought to "let the children teach themselves," which could not be farther from the truth. The best constructivist teachers also possess the highest levels of content knowledge for their discipline. This however, *can* be a serious problem in an elementary and middle school science classroom. The average elementary and middle school science teacher is a woman, and most women are uncomfortable with science as a topic to teach. I will discuss this in depth a little later, but it is worthy of discussion.

COMPONENTS OF CONSTRUCTIVISM

In order to understand constructivist practices in terms of their origins in psychology and educational philosophy, it is necessary to separate them into components, along with their corresponding research studies. The components that this book will address are concept formation, cooperative learning, alternative assessment, hands-on/active learning, and student-centered learning.

Concept Formation

Lev Vygotsky(1896-1934) was a Russian psychologist interested in the formation of language and thinking and cognitive development. Vygotsky stated that one of the basic components of constructivist pedagogy is the notion that children develop concepts on their own through everyday experience, called everyday concepts, and those concepts learned in school, called scientific concepts. These scientific concepts may be remote from a child's experience unless a teacher knows how to tie them into the child's experiences to make them meaningful. Conceptual change is the term that refers to the ongoing process in which children integrate their everyday concepts into a system of related concepts, including scientific concepts that have been taught in school (Howe & Jones, 1993).

Vygotsky, more than any other philosopher other than Dewey, had a huge impact on constructivist science classrooms. Vygotsky contended that students learn better through social interaction, since discussions and feedback take learning to higher and higher levels. Nowhere in education is this better represented than in a science lab with cooperative groups. Problem solving, hypothesizing, and scientific discovery demand different viewpoints and the cooperative group provides this to students. Also important is that in the real science world, experiments do not always lead to great discoveries. Science students must learn that there is value in narrowing down possible solutions until the correct one is found, and this concept of conceptual change is necessary to move from one idea of possible solution to another more valid solution. (Conceptual change theory will be discussed in more detail later in this chapter.)

The following instructional techniques help to accomplish this goal.

Reciprocal Learning

Ongoing dialogue between student and teacher is at the heart of constructivism and helps to prevent student misconceptions of learning. To gain new understandings from one's social environment and to become a high level thinker capable of making meaningful connections requires adopting specific intellectual skills that are modeled by competent teachers. Learning-to-learn strategies may be taught to students or discovered by students as they attempt to solve problems. Reciprocal teaching is one such strategy (Applefield et al., 2000-2001). Reciprocal learning and teaching strategy is the creation of Palinscar, David, and Brown (1984), but based on the work of Vygotsky. It is a strategy employed in order to raise reading comprehension, which includes four points:

1. Summarizing
2. Questioning
3. Clarifying
4. Predicting

The procedure consists of interactive dialogue where the teacher models the four skills, gradually letting the students take over the responsibility, while taking the role of coach. The teacher and students take turns leading a dialogue concerning sections of a text. They also take turns generating summaries and predictions and in clarifying misunderstandings in the text. The order in which the four strategies occur is not important, most teachers mold the four to the particular text being read (Palinscar et al., 1984). The goal is to encourage student regulated self-learning by helping students develop effective strategies and contextual knowledge of when to use them (Applefield et al., 2000-2001).

In research studies conducted by Palinscar et al. (1984), students increased their comprehension ability after receiving reciprocal teaching instruction, including modeling and corrective feedback on the four comprehension activities. The types of tasks selected for students included complex, real-life problem-based tasks, which emphasized conceptual understanding over memorization (Applefield et al., 2000-2001). Empirical support for reciprocal teaching technique is found in several comprehension studies (Palinscar et al., 1984, 1992), and results confirmed that the reciprocal technique can build pre-reading and comprehension skills (Andrews, 1985).

According to Palinscar, David, and Brown (1984), the goal of long-term reading instruction is not to focus on content knowledge that students to a large part already possess, but to stress comprehension-fostering strategies that extend knowledge to more areas other than reading. In a study conducted in 1984, teachers received training in reciprocal techniques for a reading class and students were measured on criterion tests comprehension, reliable maintenance over time, generalization to classroom comprehension tests, transfer to novel tasks, and standardized tests. These measures also were taken from traditional classrooms with no intervention. Reciprocal teaching techniques accounted for significant gains in each of these measures. Many of these results were replicated during a second study.

Reciprocal learning improved listening comprehension as well. In a study conducted at the primary level to determine whether reciprocal teaching would be an effective approach to improve nonreaders listening comprehension, before the administration of the treatment (recripocal teaching), pretest scores were 51% correct for the reciprocal group against 49% correct for the traditionally taught group. After treatment, posttest scores were 72% for the reciprocal group against 55% for the traditional group. Reciprocal teaching was compared to traditional basal reading instruction where both sets of students read the same text from basal readers (Palinscar, David, & Brown, 1992).

Reciprocal learning theory has as its foundation Vygotsky's learning theory. Vygotsky's unique ideas about education and socialization of children which are relevant to science teaching were developed through observing children going about their daily business of school, family, and play, and emphasized the importance of interactions with others as it fosters cognitive development. Vytogsky emphasized the role of guided learning in social contexts, which is the basis of reciprocal learning (Palinscar et al., 1992) and the bedrock component of today's constructivism. Vygotsky's contribution to constructivism has been identified with social constructivism because it emphasized the importance of social context for cognitive development.

Vygotsky's best known concept in the social context is called the zone of proximal development, which could be another term for reciprocal learning. It argues that students can, with the help of teachers and slightly more advanced students, master ideas and concepts that they could not master by themselves. He believed that

> children should have tasks set for them that are just beyond their present capability but which they can perform with guidance from a teacher or more advanced peer. He described a "zone of proximal development" (ZPD), as

> an area just beyond a child's current level of ability. (Howe & Jones, 1993, p. 31)

As a middle school science teacher, I tried to teach in this "zone" every day. I would pose difficult problems, but provide the metaphoric ladder that the students could use to find the answer. And if I saw their frustration growing after a time, I would throw them a portion of the answer until they got it. I provided a scaffold for their learning.

As mentioned, Vygotsky's concepts are aligned closely with science education. Successful science classrooms stress cooperative learning (as mine did), with laboratory experiments serving to enhance social skills and cooperation in the completion of science process and lab skills. In addition to Vygotsky, this style of teaching has at its foundation the theories of Dewey, Piaget, and Bruner (Howe & Jones, 1993). There are four general principles that are applied in any Vygotskian classroom:

1. Learning and development is a social, collaborative activity.
2. The zone of proximal development can serve as a guide for curricular and lesson planning.
3. School learning should occur in a meaningful context and not be separated from learning and knowledge children develop in the "real world."
4. Out-of-school experiences should be related to the child's school experience (Howe & Jones, 1993).

Vygotsky has filled in gaps some scholars find in Piaget's work, such as not including the importance of social dimensions and their influence on intellectual development. Vygotsky's theory suggests the inherent social nature of all humans and his work marries social with intellectual instead of divorcing the two. Socially mediated instruction as it pertains to Vygotsky is called scaffolding. The nature of scaffolding is for the teacher to provide enough support without doing the work for the student (Palinscar et al., 1992).

Albert Bandura (1977) has also studied human behavior in a social learning theory that he calls "reciprocal determinism." In this theory, human behavior influences environment and environment influences human behavior. People and environments do not function independently of each other, rather they determine each other. This is the opposite view of behaviorism which states that a stimulus always causes a response; a one-way directional relationship. Behaviorism neglects determinants of behavior caused by cognitive functioning. Social learning theory relies heavily on self-regulating capacities within the individual, thereby placing some responsibility on the person and not solely on the stimulus. In the constructivist classroom, this would have implications for students who interact and participate in their learning rather than experiences a more passive learning experience.

The Learning Cycle

Constructivism is based on the notion that students build knowledge by continually restructuring new information to fit existing concepts. The learning cycle is a conceptual-change model of instruction that is consistent with concept formation. It has several components that are similar to reciprocal learning. The three-stage model is as follows:

1. Exploration phase-teacher gives students materials and encourages exploration and questions about things dealing with new materials that they do not understand.
2. Concept introduction phase-teacher introduces and explains key concepts, may illustrate, diagram. Textual readings become more purposeful.
3. Concept application phase-teacher help the students apply the newly learned concept to new situations.

The learning cycle is based on the work of Piaget and his learning principles of mastery and self-regulation, where learners develop new reasoning patterns as they accommodate and assimilate new ideas. Students become reflective and as they practice new skills, they improve their cognition rather than their behavior as in the case of behaviorism, which is what drives the traditional teaching method (Ebenezer & Haggerty, 1999). Employing the learning cycle also clarifies students' thought processes and misconceptions. Students have the opportunities to explain and debate their ideas, thereby giving teachers good insight as to why students are arriving at certain answers or viewpoints (Bevevino, Dengel, & Adams, 1999). The learning cycle is crucial to science education because it brings the student back time and time again to the basic question; does the data support a new idea that suggests that I have to change my old beliefs? Much of the content in science is intimidating to students, and it takes teachers willing to train students to question belief systems in order to move on to the next, higher level of understanding. Much of learning can be blocked by students old belief systems holding them back.

Musheno and Lawson (1999) studied to see whether the learning cycle can be applied effectively to teach science text. High school students were randomly assigned to read either a traditional text passage or a learning-cycle passage. The students in the learning cycle group earned higher scores on concepts comprehension questions at all reading levels.

In addition to Piaget, accommodation and assimilation are also components of constructivism as defined by Fosnot (1989). During concept introduction, students may encounter realities than contradict their existing ideas. Cognitive conflict arises through group dynamics and social exchange as the learner realizes that there may be a contradiction between his or her understanding and what he or she is experiencing (Applefield et al., 2000-2001). Conceptual change theories of instruction are based on constructivist perspectives, and from this view, learning involves interactions between new and existing conceptions. Teaching is more than providing one correct view.

Conceptual change methods which include techniques such as learning cycles and students' changing conceptions have been shown to foster positive student attitudes. Heide (1998) demonstrated that students demonstrated more positive attitudes about science and implemented higher-order thinking skills as a result of constructivist-based conceptual change teaching. Everyone who has ever taught science understands how important for students to believe that they can "like" science and become proficient in it.

Constructivism states that conceptual change is the key to cognitive growth and development, and so conceptual change should become the goal for every good teacher's instruction (Applefield et al., 2000-2001). There is evidence that conceptual understanding of content is higher when students are taught in constructivist classrooms. Current research supports the advantage of conceptual learning over memorization. Constructivism has been very successful in mathematics instruction where students have historically done poorly in terms of understanding certain mathematical concepts, such as giving students relevant examples to solving analogous problems that have some connection to similar problems and prior knowledge (Chen, 1999).

Specifically, Chen (1999), conducted research concerning children's learning and transfer to determine the conditions under which and the extent to which children apply problem solutions from source to target (transfer) problems. Seventy-one children ranging in ages from 8 to 11 years old were recruited from a midsize city. Results showed that children who a learned a general schema (concept) that applied to a problem, had no difficulty answering problems that included formulae and enhanced their flexibility in solving the target problem. In contrast, children in the invariant group who did not learn the concept behind the formula, tended to be tied to the specific formula and so when asked to solve a problem requiring a different formula, they experienced difficulty solving the problems.

Of course, teacher competence can either enhance or sabotage constructivist learning experiences. Success with constructivism is dependent partly on teachers possessing sophisticated epistemologies and being properly trained in the technique. Some researchers go so far as to call traditional teaching techniques "naïve" and constructivism "sophisticated epistemology" (Howard, McGee, Swarty, & Purcell, 2000).

Teachers themselves must embrace constructivist practices during professional development. Berger (1999) showed that teachers must be given learning experiences based on the same pedagogical principles as the ones they are expected to implement with students, and that if teachers are going to teach for understanding, the teachers need to be challenged at their own level of mathematics competence. During a constructivist teacher workshop developed to enhance mathematics instruction, teachers were taught that conceptual learning proceeds to the development of structures, or big ideas that can generalize across experiences. Forty-eight teachers from around the state of Florida were chosen to participate based on geographic location and teaching assignment. The intent was to do a model of K-12 team approach that would later be replicated in each of the six regions of the state. As a result of the teacher education, students scored higher on algebra tests after focusing on the concepts. More important to the students, inquiry learning, which was employed in this study, showed to result in gains over traditional teaching methods in a wide range of students, especially with disadvantaged students deriving greater benefits.

There is evidence that conceptual understanding of content is higher when students are taught in constructivist classrooms. Current research supports the advantage of conceptual learning over memorization. Constructivism has been very successful in mathematics instruction where students have historically done poorly in terms of understanding certain mathematical concepts, such as giving students relevant examples to solving analogous problems that have some connection to similar problems and prior knowledge (Chen, 1999).

Schema Theory

The concept of new information being fitted into a knowledge paradigm that is already there is called *schema*. "A schema is a general knowledge framework that a person has about a particular topic. A schema organizes and guides perception" (Hyde, 1996, p. 58) It is due to this schema concept that true higher level comprehension can occur, not just memorization. When everything connects in the mind, memorization does not have to be relied upon as the core mode of learning. Regardless of the level of sophistication of a student's existing schema, each student existing schema, or knowledge structure, will have a profound impact on what is learned and whether or not real learning (as defined as conceptual change) occurs (Applefield et al., 2002-2001).

Schema is also about putting things into their proper context. Environments where children can interact with their peers, teachers, toys, or instructional materials, enhance their development and their desire to learn. When children play, they use their senses to experience the world; they feel, see, hear, and sometimes taste the world and the objects with which they are playing. When learning is dynamic as in this scenario, new information is placed into its proper schema or context, depending on the situation. Children develop nuances and subtleties otherwise not noticed. In this way, research has shown that too much teacher-directed instruction has either negative effects at worst or neutral effects at best on children's development (Meade, 1999). What is learned tends to be context-bound and tied into the situation in which it is learned (Lave & Wenger, 1991).

Athey defines schema as forms of thought. Athey worked with Meade on a project observing children's actions in relation to schema learning and brain development (Meade, 1999). Meade studied the effects of curriculum intervention on the richness and amount of stimulation teachers give 4-year-olds when they observe children who are fascinated by schemas. The researchers observed 20 nursery school children at play with particular schemas, described as lines, curves, and space order. Meade was interested in the study in terms of brain development and neural pathways. Results showed that the strengthening of neural pathways is enhanced by focused play, a self-organized focus on the schemas, even though adults may not see the play as beneficial. If a child showed interest in a "schema," such as being fascinated by horizontal lines that connect A to B, the teachers would give the students materials for them to connect; ribbon, string, and so forth. They did not give lessons, but simply observed the children. This "provision of diverse experiences" resulted in higher IQ scores because of the enrichment of the children's' experiences.

As a teacher, I have heard the argument from students many times concerning the "meaningfulness" of the subject matter and topic in any given class. "How will this be important to me in the future?" they ask. Yarlas (1999) argues that interest in a particular subject or class in large part depends on the usefulness and comprehensibility of the information, its meaningfulness to the student, and its ability to be processed and incorporated into a person's existing schema or knowledge structure. Thus, the degree to which information attainment leads to schema enhancement seem to be related directly to the student's interest level for that information. Of course, a student has to take ownership of the subject matter in order to do this. But a good science teacher can take relevant problems in the world today, like global warming or the spread of disease, and make it come alive in class by connecting world news to relevant classroom activities.

Much has been written lately concerning gender and interest in science as a content area. Yarlas (1999) chose physics classrooms and studied the effect gender had on cognitive interest. This was accomplished through assessing a learner's current state of knowledge in a domain, and creating materials that optimized the student's degree of schema enhancement. Students were read passages that contained information about either an expected or an unexpected outcome. Students were asked to either explain or describe information related to these outcomes. Schema enhancement was related to unexpected outcomes, thereby increasing interest. The data strongly supported the prediction that the more interesting the passage, the more learning occurred. Individual interest and gender were covariates because males naturally have more experience with physics and science in general, providing further evidence that supported the central hypothesis of the knowledge-schema theory; that learning increases interest for information in classroom situations where concepts are taught in ways that maximize interest.

Participants in Yarlas's (1999) study demonstrated greater learning for concepts that were related to their own knowledge-schema, than for concepts less related to their schema. This supported the prediction that the more relevant the new information is to existing information already in the child's brain, the more interesting is was for the child, possibly explaining why girls fair poorly in advanced physics classes. Girls start out strongly in science in elementary school, but by middle school drop out in high numbers, thus eliminating any science schema structure on which to build. More about this will be discussed in a later chapter.

Walker (1999) conducted a differential item functional analysis to determine if seventh and eighth grade students participating in the Third International Mathematics and Science Study who were taught mathematics in a constructivist classroom had a higher probability of obtaining the correct answer to mathematics items that measured conceptual, rather than procedural understanding, than students taught in a traditional classroom. Results showed that the constructivist taught students had a higher probability of answering mathematical items that measured conceptual understanding correctly, than students taught in traditional classrooms.

Bruner (1960) contends outside forces or experiences, in addition to growth and maturation, may propel a child from one stage of development into the next. As a cognitive psychologist, the fundamental assumption of Bruner's work is that humans use mental models to represent reality. These models also can be described as modes of representing knowledge and experience:

1. Enactive—from infancy, this mode corresponds to Piaget's sensorimotor stage. This representation is experience translated into action.
2. Iconic—these representations use visual imagery and develop at age two to three.
3. Symbolic—language and mathematics systems and develop from around 7 years of age.

Bruner (1960) moves into an interactionist position in his theory of learning, encompassing constructivism, and emphasizing the roles of exchange between teacher and learner in the acquisition of knowledge. He developed the notion of "the spiral curriculum" whereby the curriculum should involve the mastery of skills that lead to the mastery of higher level skills throughout a child's academic career. For example, the topic of acceleration can be taught in a simple way in first grade, to a more complex way in middle school, to a very detailed formula driven physics class in high school (Howe & Jones, 1993). The example I gave earlier about girls dropping out of middle school science illustrates how spiraling of science content cannot occur when girls drop out of classes.

According to Bruner (1960), learners construct their own meaning through concept formation, and that the learner selects and transforms information, constructs hypotheses, and makes decisions relying on mental models to do so. In order to operationalize Bruner's theories, teachers must be active problem solvers with expectations for the students to be interactive learners. Process is important to Bruner, therefore science education is the perfect vehicle with which to carry out his ideas.

Bruner's (1960) concept formation serves as a vital ingredient in the constructivist classroom. In a study conducted by Discenna and Howse (1998), 22 preservice elementary education students enrolled in either a physical science or life science course were instructed by one of the authors at a midsized Midwestern university. The researchers were seeking to enhance preservice teachers' scientific knowledge by changing their notions of science and their epistemological beliefs of on science learning. The authors were interested specifically in describing beliefs that students bring to the science classroom and to science learning as a meaning-making activity and how these beliefs in science may differ from beliefs about learning.

Both classes stressed problem solving and guided inquiry activities as the method of teaching science. During 15 weeks, the subjects participated in a guided reflection task. After the course, the journals were coded into five "views" of how science should be taught. The most passive view considered science a body of knowledge or set of facts to be memorized by listening. The more active considered science to be the replicating of work by others. A middle view depicted science as existing in objects and that in order to learn, manipulating these objects to discover the "science" behind them was important. Students' ideas changed in a positive way during the semester in terms of science learning, aiding their concept formation. The authors argue that pre-service teachers need more classes in the inquiry/problem-solving tradition with teacher mentors. When preservice teachers are trained in schema-theory, they begin to understand that the notion of science making and science learning as a meaning-making enterprise. This is very important in fostering the same traits in students once the teachers reach the classroom (Discenna & Howse, 1998).

Cooperative Learning/Social Learning

John Dewey led the way for progressive education reformers at the turn of the century. Dewey held that education was composed of four main objectives: intellectual, moral, social, and aesthetic development. The development of the whole child became the goal. Although the term "constructivism" is never to be found in Dewey literature, his philosophy is the buttress of the whole constructivist movement.

Dewey was the first philosopher to recognize the social as well as the intellectual aspects of learning. Dewey was a proponent of the "social gospel theory" of religion, which stressed the living of one's life for the greater good of others, and for workers and employers to work towards each other's best interests. He wrote of "education as a social function" whereby teaching consists of "social direction" (Dewey, 1916). Note that the role of teaching according to Dewey is not to lecture and impart knowledge, but to direct student activity to discover his or her own knowledge. The classroom consists of a "social environment" (p. 14).

Indeed, social constructivism, supports cooperative learning. According to Vygotsky, children develop in social or group settings. Instead of working alone, children benefit when the teacher serves as the guide, encouraging students to work in groups to discuss issues and challenges that are rooted in real life situations. Teachers thereby facilitate cognitive growth and learning, as do their peers (Anonymous, 1997).

Cooperative learning is based on the Deweyan notion of social learning. Science is the perfect curriculum area for the employment of cooperative learning, since the very nature of scientific exploration includes social learning between laboratory partners. Much empirical evidence exists suggesting that cooperative learning enhances not only a more thorough mastery of skills, but also social and communication skills as well (Slavin, 1985). In sum, Dewey reformulated the framework for education, by stating that learners make sense of new information by placing it in already existing schema, a basic part of constructivism. He dramatically influenced education, and continues to have great presence in the educational arena.

Cognition is viewed as a collaborative process and constructivist thought provides a theoretical basis for cooperative learning, which points toward the powerful social aspect of learning. Students are exposed to their peers' thought processes and opposing views. Constructivists also make use of cooperative learning tasks in relation to learning and comprehension, as well as peer tutoring. Students learn best in situations where they dialog with each other about problems (Applefield et al., 2000-2001).

Johnson and Johnson (1994) state that the effectiveness of cooperative learning has been confirmed by both demonstration and theoretical research. Achievement is greater when learning situations are structured cooperatively rather than competitively or individualistically; students focus both on increasing their own achievement and that of their groupmates. Cooperative learning experiences promote greater critical thinking skills, more positive attitudes about science, greater collaboration skills, better psychological health, and greater perceptions of the grading system as being fair.

Students' notions about science also are affected by cooperative learning. Science is learned by doing and is interwoven with problem-solving activities aimed at involving students in the concepts of science, as well as the pursuit of the scientific method. Teachers have the power to incorporate cooperative learning into their classrooms. According to Yager (1997), principal investigator for the Salish Project, teachers who hold student-centered beliefs were likely to have completed teacher-education programs in which they participated in cooperative learning themselves.

A number of positive outcomes have been attributed to cooperative groups, especially among girls, which will be discussed in more depth in chapter 7. When done correctly, cooperative learning is designed to reduce competitiveness while increasing cooperative spirit, heterogeneous and racial relations, and boosting academic achievement. Teachers must be aware of potential problems with cross-gender cooperative groups because boys can tend to become dominant in the group and suppress the girls' learning (Bailey, 1992).

In a study done by Slavin (1985), 504 mathematics students in Grades 3, 4, and 5 in a suburban Maryland school district were assigned randomly to one of three conditions: Team assisted individualization (cooperative groups), individualized instruction, or without student teams, or control (this group used traditional methods). These treatments were implemented for 8 weeks in Spring, 1981 to evaluate the effects of cooperative learning on achievement, attitudes and behaviors of the students. The cooperative groups gained significantly in achievement than the control group. The results on the "liking of math" scale showed indicated a significant overall treatment effect. Statistically significant overall treatment effects were found for all four of the behavioral rating scales. Six more experiments were conducted. In each of these, classes using cooperative groups were compared to untreated control classes on a variety of dependent measures. In five of the six studies, achievement in the cooperative classes was significantly higher than in the control classes.

Heide (1998) has demonstrated that students' attitudes towards science are more positive when they engage in behaviors such as choosing problems and finding solutions to those problems (student-centered), working in large and small cooperative groups, performing hands-on science laboratory experiences and learning through conceptual understanding rather than memorization.

Alternative Assessment

As was mentioned in chapter 1, assessment should match instruction. When teachers teach mostly knowledge level fact-based curricula, they assess this way also. The problem lies in how to properly assess students who are learning at higher levels in more constructivist based classrooms. There is no way around the fact that higher-level learning cannot be properly assessed by lower-level assessments, yet that is what we do when we assess the standards using high-stakes multiple-choice tests. Among the most important aspects of teaching is reaching agreement on how to determine if the learner can demonstrate in some fashion the desired learning outcome or performance (Applefield et al., 2000-2001).

There are several ways to operationalize ideas about teaching at higher levels. The first is to employ Benjamin Bloom's taxonomy of cognitive levels. In 1956, Bloom developed a brilliant classification system that still stands up today, whereby intellectual behavior important to learning was separated into three domains: cognitive, psychomotor, and affective. The cognitive domain was further divided into six levels, which demonstrate different intellectual skills. These go from the lowest levels of learning to the highest. (Verb examples are included that represent measurable intellectual activity: these ingeniously are used to write daily objectives for lesson plans, since verbs are action, and we can measure (assess) them).

1. Knowledge (lowest level) arrange, define, duplicate, label, list, memorize, name, order, recognize, relate, recall, repeat, reproduce.
2. Comprehension: classify, describe, discuss, explain, express, identify, indicate, locate, recognize, report, restate, review, select, translate.

3. Application: apply, choose, demonstrate, dramatize, employ, illustrate, interpret, operate, practice, schedule, sketch, solve, use, write.
4. Analysis: analyze, appraise, calculate, categorize, compare, contrast, criticize, differentiate, discriminate, distinguish, examine, experiment, question, test.
5. Synthesis: arrange, assemble, collect, compose, construct, create, design, develop, formulate, manage, organize, plan, prepare, propose, set up, write.
6. Evaluation: (highest level) appraise, argue, assess, attach, choose compare, defend estimate, judge, predict, rate, core, select, support, value, evaluate (Bloom, 1956, pp. 204-205).

Basically, teachers employing constructivist techniques teach at higher levels than are found in traditional (non-constructivist) classrooms. Lower level instruction is very easy and cheap to evaluate and assess, namely multiple-choice, true/false tests.

Although historically, students taught in both traditional and constructivist classrooms may or may not score similarly on multiple-choice tests, in questions dealing with comprehension, constructivist-taught students had the edge. In a paper presented to the American Educational Research Association conference, it was reported that middle school students who were taught math in a more student-centered conceptual way, had a higher probability of obtaining the correct answer to mathematics items that measured conceptual rather than procedural understanding. The students in this study were 13-year old seventh and eighth graders, who participated in the Third International Mathematics and Science Study (TIMMS). They were administered multiple-choice mathematics items from the TIMMS test as the measure of mathematics ablity. Performance expectations included knowing, using routine procedures, reasoning, and communication. Content areas covered fractions, number sense, algebra, data representation, and analysis and probability. A variant of matrix sampling was used in the test design. Differential item function analysis was used to analyze the data.

Results measured more of a conceptual understanding of mathematics and also a gain for students taught in a more student-centered environment. The students tested also were more successful in obtaining the correct answer to mathematics items that measured conceptual, rather than procedural understanding. According to Walker (1999), students should have acquired a conceptual understanding of the mathematics being taught, knowing not only *what* to do but *why* they were doing it. The conceptual understanding acquired by these students should enable them to apply their knowledge in new mathematical situations.

In a study that examined teachers' student learning outcome goals and their corresponding assessment practices, Bol and Strage (1996) note that although the national trend is toward integrating the science curriculum into students' daily life aimed at conceptual understanding rather than memory of content, teachers' assessment styles show little correspondence between these goals and actual teaching practices. In fact, teacher developed classroom tests contain mostly low-level questions in terms measuring knowledge. Although teachers' instructional goals were meant to promote higher order thinking skills, the test items included on their assessments do not reinforce those goals. Research also has shown that science teachers stress memorization over conceptual understanding (Gallagher, 1991), thereby reinforcing the need for multiple-choice assessments. The mathematics included in science intimidates many students. Not only do science students have to memorize mathematical formulae, they are then asked to grasp difficult scientific theory in application of the concepts. The problem lies in assessing the higher levels of learning, where memorization is the lowest (Bloom, 1956).

Alternative types of assessment (also called authentic assessment) can be compared and contrasted to more traditional assessment practices that would include standardized tests that feature closed-ended questions. Scores on standardized tests reflect whether or not a student selected the correct answer, but do not reflect the level of comprehension or problem-solving strategies used to arrive at the answer. Bol, Stephenson, and O'Connell (1998) conducted a study where 893 teachers in a large mid-Western urban school district were surveyed to determine assessment practices and their perceptions concerning their practices. Data were analyzed using ANOVAs, and results showed that among teachers in the field, elementary teachers are more likely to use alternative assessment methods than higher-grade teachers, and math teachers reported employing alternative assessment more frequently than did science and social studies teachers.

According to Shepard (2000), a broader range of assessment instruments is needed to measure learning goals and processes and to connect assessment directly to ongoing instruction. While multiple choice standardized tests are appropriate for measuring certain levels of acquired knowledge, Shepard suggests more open-ended performance tasks for measuring higher level thinking skills. Not only do teacher made tests measure low-level thinking skills, so do state and district tests. Statewide accountability tests (such as the Standards of Learning in Virginia), used to measure basic knowledge in science, have been corrupted with a heavy-handed rewards and punishment system doled out by administrators who do not reward the excitement of ideas. Types of alternative assessment that would ensure the proper measurement of higher-order thinking would include both informal and formal assessment tools. Some less formal evaluations would include feedback from teacher to student, dynamic ongoing assessment instead of a one-shot final test grade, self-assessment, and teacher assessment. More formal would include portfolios, rubrics, and performance-based assessment. *Assessment for learning must overcome assessment for passing tests.*

According to Gega and Peters (1998), these alternative assessment tools successfully measure higher-order thinking skills:

1. Performance—based assessment-models based on scientific concepts, experiments, journals, written material including papers.
2. Projects—requires self-assessment from start to finish. Students display critical thinking, persistence, inventiveness, and curiosity.
3. Peer or self-designed instruments—rubrics, surveys. Promotes independence and ownership.
4. Interviews—are effective ways of gaining information with students with writing problems or with very early elementary aged students who cannot express themselves in writing.
5. Journals—useful ways to get students to write to learn.
6. Portfolios—a sample of work collected over time, a good self-assessment tool.
7. Concept maps—organizes thoughts and concepts. Helps to see how things are connected, including old and new information.
8. Teacher observations—an informal, on-going tool that puts learning in context.
9. Questioning techniques—open ended questions where there is more than one correct response.

Student-Centered Learning

In discussing the nature of science, Clough (2000) argues that significant consensus exists regarding many issues appropriate for middle and high school students. Some of the most important of these ideas for helping students better understand the nature of science include: science is not the same as technology, a universal, historical scientific method does not exist, science is not completely objective, knowledge is not democratic, words used in science may not mean what students think they do, science is bounded, anomalies do not always result in rejection of an idea, scientific thinking often departs from everyday thinking. Clough suggests that students' understanding is woven into the fabric of their prior experience, which is useful in helping them make sense of new experiences.

The old out-of-date (traditional) trend in middle school science education was the idea that teaching is the transmission of discrete facts, pieces of information and specific processes. The current trend is a broader, more holistic approach that encompasses several areas of instruction, which, in turn, enhances students' understanding and comprehension. Among these are: concepts, processes, applications, attitudes, creativity, and the nature of science. When science instruction focuses solely on transmission of information, only two domains of science are addressed; concepts and processes. Students are presented with a very restricted view of science. This holistic approach develops higher levels of understanding and enables students to "do" science themselves (Daas, 2000). The Western view of the classroom has held that the student is the receiver, not a producer of information. The teacher is idealized as the ultimate source of knowledge and as a highly efficient manager (de Esteban & Penrod, 2000). In a constructivist classroom environment, the teacher's role changes to one of guiding rather than telling the learner the information (Applefield et al., 2000-2001).

For much of the twentieth century, teachers sought to teach facts in a lecture format to students. Now, educators know that teaching children how to think, solve problems and process information is more important than teaching them to memorize facts. Taba adhered to the Deweyan philosophy of education, and agreed with his brand of empiricism (pragmatic instrumentalism) in which "facts" are used to illustrate ideas and not the other way around. Taba believed that teaching should be organized through key concepts, where content should not be seen to dominate any chosen instructional method (Guyver, 1999).

Taba (1932) posits that teachers rely too much on subject matter, forcing them to decide which content to include and exclude by the end of the school year, although she warns against going too far in either direction, stating,

> As a result of a strong reaction against the emphasis on subject matter found in the traditional type of school, progressive education has regarded the child too much as a psychological phenomenon, failing to realize fully that the experience of the child is a product of its contact with the objective materials of its environment. Instead of subject matter alone doing it, the child only is now dictating educational procedure. (pp. 251-252)

Taba's response to increasing the knowledge base is to emphasize the "acquisition, understanding, and use of ideas and concepts rather than facts alone." This reduces the amount of detail to be covered in class, and it provides better conceptual links between pieces of factual information. Broader categories of knowledge like concepts, generalizations and conclusions act to impose structure on factual bits of information, linking these specific bits in categories so that a large amount of specific detail is subsumed within a limited number of ideas. Of course, in our age of standards, teachers have less leeway in this area, and have to teach certain concepts according to the state standards.

Taba (1932) developed a model to categorize information. It is a multipurpose approach that provides an occasional teaching option. The method involved three stages:

1. Students make an exhaustive list of observations, ideas, or concepts.
2. Students gather all similar items together.
3. Students name each category. They then are assigned to category groups and proceed to research their topic. The teacher's role is to facilitate acquisition of relevant information sources. The final product is a report, portfolio, project, or video presentation (Armstrong, 1998).

Taba (1966) also writes that conventional instruction does not reach those adolescents with cultural and educational deficits and that traditional instruction does not meet the needs of these students, because it is incompatible with the needs of those students. Unfortunately, most students today who have educational deficits are poor, minority, and urban, due in part to lower teacher expectations, lack of educational support at home, and lack of funding for poorer schools. Socioeconomic status is the best predictor of both grades and test scores (Bailey, 1992; Wilson, 1997).

Student-centered classrooms have been the topic of empirical study as well. In a study conducted by Chang (1994), constructivist, student-centered classrooms produced students who scored much higher when asked to explain certain scientific phenomena, than students in traditional classrooms. A sample of 363 eighth grade students in a junior high school in Taipei, Taiwan were divided into three groups. All groups were given multiple choice tests, and the scores on the tests were similar in both groups. The difference showed up in the comprehension (as evidenced by explanations) of the subject matter. Also, a teacher main effect appeared in the results of the $3 \times 2 \times 2$ and 3×3 ANOVA analysis, indicating that teachers made significant differences on students' posttest scores. However, results indicate that teacher characteristics, more so than teaching technique, contributed to the results.

Dunkhase, Hand, Shymansky, and Yore (1997) conducted a study comparing a more student-centered environment against a teacher-centered environment and outcomes concerning student attitudes and perceptions. The study focused on student perceptions of their science instruction and student attitudes toward science learning as a function of their exposure to interactive-constructivist teaching strategies aimed at student ideas, utilization of literature integration, and incorporating parents as partners. Among the components of the student-centered environment were interactive-constructivist teaching strategies designed to focus on student ideas, shared control, listening to students' ideas, and making ideas and practices meaningful at the individual student level. Two groups were designated, students from classsrooms where teachers were instructed in constructivist philosophies, and students from classrooms without such instruction. The results showed that attitudes and perceptions were higher in the constructivist/student-centered classrooms than in the traditional classrooms. Of note is the fact that girls experienced the highest rise in attitudes and perceptions concerning the teacher delivery approach, while boys experienced a rise in positive attitudes concerning content.

Active learning includes student participation. Participation encourages students to exchange ideas and viewpoints freely in order to clarify, evaluate, and reconstruct existing schema. In fact, the very effectiveness of a constructivist approach depends on students actively participating in classroom activities. Research shows that constructivist classrooms can increase students' ability to reconstruct their knowledge and that students in constructivist classrooms are challenged to be more active learners (Applefield, 2000-2001; Thomasini & Others, 1990).

Research also shows that students learn more when they have some ownership in the learning process; the basis of constructivism. Yager (1997) states that science students viewed science as more relevant to their daily lives than mathematics was to mathematics students, and that new teachers recently graduated from teacher colleges saw themselves and their classrooms as very student-centered. This study also states that more teachers think they are student-centered when actually their classes are teacher centered, however, students who behaved in student-centered ways were taught by new teachers who held coherent student-centered philosophies of teaching.

Yager et al. (1997) also showed that teacher education programs are crucial for teachers who want to be student-centered in philosophy. Among the findings in this area are these:

- Student-centered actions were not observed in classes taught by new teachers whose philosophies of teaching were not coherent with their practices.
- Students who behaved in student-centered ways were taught by new teachers who held a coherent student-centered philosophy of learning.
- New teachers holding student-centered beliefs were likely to have completed teacher preparation programs were they engaged in

cooperative learning, were assessed of their performance in the field, and had strong, close personal relationships with faculty.

- Student-centered teachers were more likely to have completed a longer student teaching experience.

Developmental Stages/Readiness

Jean Piaget made huge contributions towards our current knowledge of intellectual and cognitive development. Piaget brought to light the constructivist notion of readiness; or how children learn in relation to what stage of development they are in currently. Constructivism states that children bring different levels of abstraction, knowledge and understanding to every learning experience, based on cognitive readiness. This concept is where the child-centered constructivism component developed.

The Victorian notion of children as miniature adults was smashed in part by Piaget and his radical views on child development. Piaget's stages of cognitive development are listed below:

1. Sensory Motor—birth to 2 years. Babies learn to use their bodies and movement to connect with the outside world. The baby begins to understand how one thing can affect another, and simple notions about time and space.
2. Preoperational—2 to 7 years. Egocentric stage, babies can only consider their point of view. "De-centering" occurs during this stage, where the child comes to realize that there are other people in the world and that the world does not revolve around them. Not yet logical thinking ... children can fear going down the drain in the bathtub or falling through a crack in the walkway of a fishing pier.
3. Concrete operational—7 to 11. Thoughts become more rational and operational. Thoughts are formalized around concrete objects in the presence of the child.
4. Formal operations—11-16. Thoughts are formalized through hypothetical processes, objects do not have to be present. Abstractions are possible.

One can see the influences of Dewey (1916) in these stages of cognitive development. Dewey states:

> The aim of natural development translates into the aim of respect for physical mobility ... (and that) we find natural development as an aim, enables him to point the means of correcting many evils in current practices, and to indicate a number of desirable specific aims. (p. 115)

Constructivist Learning Environment

Since constructivism does not tell the teacher what former experiences students should have, it does caution teachers against instructional techniques that may limit student understanding. Knowledge is not objective, but the teacher organizes information around conceptual clusters of problems, questions and discrepant situations in order to engage the student's interest (Hanley, 1994).

Driver (1989) has identified certain features that should be present when science is taught from Constructivist pedagogy:

1. Identify and build on the knowledge that learners bring to the lesson.
2. Allow the learners to develop and restructure this knowledge through experiences, discussions, and the teacher's help.
3. Enable pupils to construct for themselves and to use appropriate science concepts.
4. Encourage pupils to take responsibility for their own learning.
5. Help pupils develop understanding of the nature of scientific knowledge, including how the claims of science are validated and how these may change over time (pp. 83-106).

Brooks and Brooks (1993) pose the following as their description of a constructivist classroom setting:

1. They free students from the boredom of fact-driven curriculums and allow focus on large ideas.
2. They turn over to the students the power to follow trails of interest, to make connections, to reformulate ideas, and to reach unique conclusions.
3. They share with students the important message that the world is a complex place in which multiple perspectives exist, and truth is often a matter of interpretation.
4. They acknowledge that learning, and the process of assessment, are elusive and messy endeavors that are not easily managed (p. 32).

To date, many researchers have proposed models of ideal classroom environments. Excellent science classrooms are managed by teachers who use strategies that facilitate sustained student engagement, increase student understanding and comprehension of concepts and scientific knowledge, and encourage student participation in an active learning environment. A recent study suggests that there are advantages of participatory classroom environments, where students construct their own sensory input and make inferences in that information to draw conclusions (Strage & Bol, 1996).

Constructivist classrooms foster communication between students and teachers and among students themselves. Communication apprehension (CA) or "fear to communicate" was studied as a response to teacher philosophies in the classroom. A purposefully selected sample of 61 student teachers during their education program were given the personal report on CA to identify their levels of CA. A Pearson *r* was used to analyze the data. Results showed that high levels of CA are related to nonconstructivist previous school experience. These people are assumed to have had experienced more traditional teaching styles by their teachers while in school (de Esteban & Penrod, 2000).

Howe and Jones (1993) offer this outline of the major contributors to the constructivist movement and their implication for the science classroom (see Table 2.1). Bowers (1991) offers the following instructional model operationalizing constructivism (see Table 2.2).

DEVELOPMENT OF DIRECT INSTRUCTION

Direct Instruction

Traditional instructional technique is the current instructional strategy based on this philosophy and is based on 100 years of research. The term "direct instruction" was coined by Engelmann. It is the pedagogy currently driving the standardized test movement. From 1966 to 1969, Engelmann was involved in a number of grant-funded projects aimed at exploring the extent to which special instructional methods and innovative curricular approaches would enhance the learning of children. It was during this period that Engelmann coined the term "direct instruction" and formalized the logic and methods for the operationalization of this instructional method. Engelmann's early work focused on beginning reading, language, and math. It was published by Science Research Associates in 1968 under the trade name DISTAR (Direct Instruction System for Teaching and Remediation). Over the past 3 decades, the original curricula have been revised and new ones developed. These curricula have been incorporated into the comprehensive school reform model known as the direct instruction model, which has been implemented in some 150 schools nationwide (Parsons & Polson, 1931).

There are several working definitions for direct instruction. Direct instruction is described by McDermott (1993) in this way: "Instruction in introductory physics has traditionally been based on the instructor's view of the subject and instructor's perception of the student." The teachers in

Table 2.1. Constructivism in Science Teaching

Scholar	*Major Ideas or Themes*
Piaget	Children acquire knowledge by acting and thinking. Knowledge is classified as physical, logico-mathimatical, or social. Development of logical thinking is a maturational process. Understanding of natural phenomena depends on logical thinking ability.
Bruner	Children learn by discovering their own solutions to open-ended problems. Knowledge is represented in enactive, iconic, and symbolic modes. Appropriate ways can be found to introduce children to any topic at any age. The process of learning is more important than the product.
Vygotsky	Children learn through interaction with peers and adults. Knowledge is built as a result of both biological and social forces. Language is a crucial factor in thinking and learning. Children need tasks just above their current level of competence.
Kohlberg	Children learn moral and ethical behavior by example rather than by teaching. Moral development is a slow, maturational process. Moral dilemmas that have no easy solution are part of life. Learning is domain-specific. Misconceptions about natural phenomena interfere with new learning. Both procedural and declarative knowledge are important.
	Implications for Science Teaching
Piaget	Provide environment to encourage independent action and thought. Distinguish between kinds of knowledge in planning instruction. Be aware of children's level of thinking.
Bruner	Use open-ended problems in science regularly and often. Use all three models of teaching and testing for understanding. Emphasize processes of science. Teach concepts and processes that will lead to further learning.
Vygotsky	Encourage pupils to work together and to learn from each other. Encourage children to explain what they are doing and thinking in science. Set tasks that challenge children to go beyond present accomplishment.
Papert	Make sure that children understand the meaning of their class activities. Make the computer a tool for new learning, not a substitute for a book. Encourage and model thinking about thinking.

Source: Howe and Jones (1993).

this scenario are eager to transmit their knowledge to the student. Generalizations often are formulated upon introduction, and students are not actively engaged in the process of abstraction and generalization. The reasoning is almost entirely deductive, very little inductive thinking is involved. McDermott states:

> The trouble with the traditional approach is that it ignores the possibility that the perception of students may be very different from that of the instructor (as is the case in constructivist philosophy). Perhaps most students

Table 2.2. Operationalizing Constructivism

	Concept Formation	*Exploration*
Introductory	*Students use descriptive science processes	*Hands on experiences
	Data Interpretation	*Conceptual*
Developmental	*Students identify and investigate relationships	*Student discussion groups
		*Teacher-directed discussions
	*Make inferences	*Students form concepts
	*Use integrated science	*Comparison of student concepts
	processes	with expert concepts
	Application of principles	Application
Culminating	*Students make predictions and hypotheses	*Concept expansion
	*Support and justify predictions and hypotheses	*Investigation of science/technology/ society issues
	*Test predictions and hypotheses	
	*Verify predictions and hypotheses	

Source: Bowers (1991).

are not ready or able to learn physics in the way that the subject is usually taught.

In contrast to supporters of constructivism, proponents of direct instruction believe that:

1. External reality does exist independently of the observer.
2. Humans have organized knowledge into systems to better understand reality: such as mathematics, biology, literature, and history among others. The role of teachers is to help students acquire this knowledge.
3. Direct instruction proponents believe that educators are guided by the main concepts of "behavior" and "learning." Behavior is anything students do that is observable. However, direct instruction also cares about how students feel, think, and act.

4. The second main concept is learning, defined as a change in behavior that results in direct interaction with the environment, that is, from teaching-systematic or incidental (Kazloff, LaNunziata, & Cowardin, 1999).

Marchland-Martella, Martella, and Lignugaria-Kraft (1997) developed a system to use while observing practicum teachers delivering a direct instruction lesson (see Table 2.3).

Table 2.3. Definitions of Correct Direct Instruction Behaviors

Presentation—

Cue—Focus word, phrase, or question (e.g., what word?, get ready) as indicated by program format or as specified by teacher.

Pause—At least a 1 second waiting time (preferably 2 seconds)

Signal—Hand, touch, or auditory response presented by teacher which initiates a pupil response

Responses—

Group—Two or more pupils respond simultaneously and correctly

Individual—Pupil responds correctly

Signal Error Corrections—

Address—Corrects within 3 seconds after group error occurs; addressed to group; positive tone (without negative comments or gestures); tells group what they have to do (e.g., *I've got to hear everyone. You have to wait until I signal*)

Repeat—Repeat original presentation to test group's response; positive tone (without negative comments or gestures)

Response Error Corrections—

Model—Corrects error within 3 seconds after group/individual error occurs; addresses model to group (if group response) or individual (if individual response); positive tone (without negative comments or gestures); demonstrates correct response to pupil(s).

Test—Requests group/individual to respond again using original cue provided before error occurred; addresses test to individual if individual response or group if group response; positive tone (without negative comments or gestures)

Praise Statements—

Specific—Precise statement that reflects a positive response to a desired behavior (e.g., *Nice job saying* brother) which is delivered after an appropriate behavioral or academic response (e.g., pupil is sitting quietly with hands folded).

General—Global or broad statement that reflects a positive response to a desired behavior (e.g., Super) which is delivered after an appropriate behavioral or academic response (e.g., student completes homework assignment)

Source: Marchland-Martella et al. (1997).

As is shown, direct instruction relies on teacher-centered lecture, with students having one correct answer. According to McDermott (1993), the reason many students do not understand subjects like physics is that the teachers rely solely on transmitting knowledge from themselves to their students, and that the trouble with the traditional approach to instruction is that it ignores the possibility that the students may have a different perception of the subject than the teacher has. Most science teachers view their students as mini-versions of themselves, when that is not the case.

McDermott (1993) also offers these shortcomings of traditional instruction:

1. Facility in solving quantitative problems is not an adequate criterion for understanding. Questions that require qualitative reasoning and verbal explanation are essential.
2. A coherent conceptual framework is not usually the outcome of traditional instruction: Students must participate in the process of constructing qualitative models that can help them understand relationships and differences among concepts.
3. Certain conceptual difficulties are not overcome by traditional instruction. Persistent conceptual difficulties must be addressed by repeated exposures in more than one context.
4. Growth in reasoning ability does not result from traditional instruction and scientific reasoning skills must be cultivated.
5. Connections among concepts, formal representations, and the real world are lacking after traditional instruction. Students need practice in interpreting physics formalism and relating it to the real world.

Teaching by telling is an ineffective mode of instruction for most students. Students must be intellectually active to develop a functional understanding (McDermott, 1993, p. 105).

Another term used for direct instruction is "instructivist" approach, a term coined by Finn and Ravitch in 1996 in their report *Education Reform 1995-1996: A Report from the Educational Excellence Network to its Education Policy Committee and the American People*. Finn and Ravitch argue that constructivism is faddish and that it excludes content. In a paragraph headed "The Romance of 'Natural Learning,' " they posit that constructivism is "hostile to standards, assessments and accountability" (p. 106).

Finn and Ravitch (1996) also argue that too much constructivism means kids who can neither read nor write, although they may have curiosity and self-esteem. Although keenly pro instructivist, they also argue for a balance in the classroom. The best teachers are not a slave to dogma, they are able to employ constructivist and instructivist techniques as the situation and child require.

In his book *Cultural Literacy,* Hirsch wrote that a content-based curricula was preferred, which ran counter to progressive educators' beliefs that natural development, process and critical thinking skills were goals to be met by education. For Hirsch (as cited in M. Berube, 1994, pp. 1-11), the fault with American education lay with the theories of Rousseau whose ideas influenced John Dewey, claiming that Dewey advocated the content-neutral curriculum.

Advanced Organizers

Ausubel's contribution to learning theory includes his belief that humans acquire meaningful learning through an interaction of newly learned information with relevant existing ideas in cognitive structure. Ausubel explored the process of what he calls meaningful learning and how it relates to a learner's cognitive structure. His "theory of meaningful verbal learning" was unveiled in his 1963 book *The Psychology of Meaningful Learning.* He also promotes the arrangement of school curriculum to match student readiness, which shows influence of Piaget.

Although Ausubel (1963) openly supports direct instruction, he also writes that the learner must make an intellectual link between newly learned information and that previously stored in his or her cognitive structure. Because of this connection, retention is greater and understanding is significant.

In order to facilitate new learning, Ausubel (1963) advocates advanced organizers; outlines of material yet to be learned, a type of summary of material that highlights key concepts and propositions for the students. Knowing that the brain builds knowledge in a hierarchical structure and by assimilating new knowledge with the help of advanced organizers, the learner builds anchors for future knowledge.

There is empirical support for direct instruction. In a study supporting traditional methods, 138 students (including 23 mildly handicapped students) in Grades 4 through 6 participated in a study aimed at comparing the effectiveness of two teaching techniques (direct instruction versus discovery teaching) in three elementary schools in a suburban Chicago school system on achievement. Students were randomly assigned to one of two treatments: direct instruction or discovery teaching. A 2 × 5 factorial design was employed. Results showed that students in both groups learned equally well as measured by a posttest. However, students in the discovery treatment group outperformed their direct instruction peers on a delayed posttest administered 2 weeks after the treatment ended (Bay & Others, 1992).

Project Head Start, a grant funded by the U.S. Department of Education between 1969 and 1972, was directed by Englemann. The purpose of the grant was to provide a comparison of the different models of educational programs for disadvantaged children. Children in three Engelmann-Becker models were compared with children in other models of instruction. This was called the largest controlled comparative study of teaching methods in history. The Engelmann-Becker model worked with twenty school districts to implement effective instructional programs in Grades 1 through 3 as part of Head Start. Research focused on specific variables that made a difference in student performance. Results showed that students in direct instruction classrooms had placed first in reading, math, spelling and language. Even though no other model was as effective, Direct Instruction has been spurned by the majority of the educational establishment (Anonymous, 2000).

Direct instruction advocates posit that behavior is anything students do, and therefore, learning is a change in behavior (feeling, thinking, acting) that results from interaction with their environment. The instructivist approach in education means that educators draw on literature on how students learn to design appropriate curricula, and focus on changes in students' behavior (learning) as a way of tracking progress (Applefield, 2000-2001).

Operationalizing Instructivism

According to Kozloff, LaNunziata, and Cowardin (1999). there are basically three distinct approaches to teaching using the instructivist method:

1. *Applied behavior analysis:* (Kozloff et al., 1999; Kozloff, LaNunziata, Cowardin, & Bessellieu, 2000) The first branch of instructivist technique is really a combination of practices derived from years of experimental research on how environmental events and arrangements affect learning and principles of operant learning, found in the work of B. F. Skinner. These tenants are as follows:

    (a) Methods for examining the interaction of students with their environments so that relationships may be discovered, that is, one can find out how a student's learning is helped or hurt by such things as difficulty, pacing, and assistance from the teacher, or the nature of their interaction with peers.

    (b) Guidelines for using knowledge of functional relationships between environmental features and a student's learning, to design instruction that is consistent with a student's skills.

(c) Methods of evaluating the adequacy of curriculum and instruction by tracking students' learning, and revising curriculum and instruction accordingly.

2. *Precision teaching:* Developed by Ogden Lindsey and associates. Lindsey based precision teaching on Skinner's discovery that the rate of behavior (# of occurrences/time) is a dimension of behavior, and not just a measure of the behavior. This implies a difference in fluent versus nonfluent behavior. The following are features:

   (a) Teachers identify and teach the "tool skills" (component or elemental skills and knowledge) needed to learn complex skills and knowledge. For example, listening to a teacher, taking notes, having fluency with math facts, etc. When students are not fluent with tool skills (reading and writing), they are not able to learn complex skills.

   (b) Teachers provide carefully planned, short practice sessions on older and new learning to strengthen retention.

   (c) As students master component skills, teachers help students to assemble component skills into complex activities.

   (d) Teachers help students keep track of their own progress.

3. *Direct instruction:* (Adams & Englemann, 1996). This third branch of the instructivist approach grew out of the work of Englemann and his work with disadvantaged children. Direct instruction was compared with 12 other methods of instruction during the largest educational study ever conducted and results showed that direct instruction was superior in fostering reading and math skills, higher-order cognitive skills, and self-esteem.

   (a) Direct Instruction focuses on cognitive learning- concepts, propositions, strategies and operations.

   (b) Curriculum development involves three analyses: knowledge, communication and student behavior.

   (c) Instruction teaches concepts, strategies and operations to greater mastery and generality. Direct instruction focuses on big ideas.

(d) Concepts are not taught in isolation from each other.

(f) The analysis of knowledge is used to create student-teacher communication.

(g) Lessons are arranged logically so that students first learn what is needed to grasp later concepts.

(h) Lessons are formatted so teachers know what to say and what to ask that enables students to reveal understanding and/or difficulties.

(i) Lessons are followed by independent and small group activity.

(j) Gradually, instruction moves from teacher guided to more student guided.

(k) Short proficiency tests are used about every ten lessons (Kozloff et al., 2001).

Bowers (1991) offers a way to differentiate between constructivism and more traditional direct instruction teaching techniques in actual classroom situations. Bowers cites Tickle who writes that the core teaching issue in middle school is the tension between the two instructional approaches; as he puts it, "one emphasizes the mastery of skills in content and the other stresses providing for the developmental needs of young adolescents" (pp. 4-9). Bowers also argues for a non-content-area-specific learning approach that would emphasize the whole child and not just rote memorization.

In differentiating between the two methods of teaching, Bowers (1991) includes examples of behaviors that would occur during each educational experience. Representing constructivism, Bowers has combined the inductive thinking theory set forth by Taba, and the learning cycle, which began several years ago as part of the Science Curriculum Improvement Study. Barman (1989) has modified the terminology of the learning cycle to make it more meaningful for elementary school teachers. Representing the traditional or direct instruction approach, Bowers (1991) cites Ausubel's (1963) "advanced organizer." The following sets of behaviors are grouped as: (1) **Introductory**; the beginning of the daily lesson, (2) **Developmental**; the operationalizing of the lesson, and (3) **Culminating**; the summation of the daily lesson. I have combined inductive thinking and learning cycle behaviors to represent the functions of a constructivist classroom and the advanced organizer for the direct instruction classroom.

Bowers (1991) offers in Table 2.4 the instructional model operationalizing direct instruction.

A paradigm describing traditional versus constructivist classroom environments is provided by de Esteban and Penrod (2000) (See Table 2.5).

In trying to bring all of this research into focus, one needs only to remember that there is a difference in techniques between constructivist teachers and more traditional teachers, and that both constructivist and traditional pedagogies alike, have meaningful and valuable components. Since this book focuses on assessment rather than instruction, it is important nonetheless to view instructional practices with assessment in mind, since we cannot have one without the other. Traditional and progressive teaching techniques must by their very nature result in different assessment strategies.

Progressive education seeks to educate the whole child, hence the use of constructivist strategies that the movement has adopted. It is important for children to *like* school, and to be engaged in relevant, meaningful tasks that force children to think, instead of merely parroting back information. A progressive notion indeed.

Table 2.4. Operationalizing Direct Instruction

Introductory
*Clarify objectives
*Give examples
*Define context
*Prompt learner's prior knowledge and experience
Developmental
*Directed teaching
*Organization of tasks
*Logical order of material
Culminating
*Students integrate new learning and prior knowledge
*The teacher promotes logical and critical approach to information
*Students resolve conflicting information and misconceptions

Source: Bowers (1991).

Table 2.5. Traditional Versus Constructivist Classroom Environment

Traditional Classroom	*Constructivist Classroom*
Curriculum is presented part to whole with emphasis in basic skills. Strict adherence to fixed curriculum is highly valued.	Curriculum is presented whole to part with emphasis on big concepts.
Curricular activities rely heavily on textbooks and notebooks.	Pursuit of student questions is highly valued.
Students are viewed as black slates onto which information is etched by the teacher.	Curricular activities rely heavily on primary sources of data and manipulative materials.
Teachers generally behave in a didactic manner, disseminating information to students.	Students are viewed as thinkers with emerging theories about the world.
Teachers seek the correct answer to validate students' learning.	Teachers generally behave in an interactive manner mediating the environment for the students.
Assessment of student learning is viewed as separate from teaching and occur almost entirely through testing.	Teachers seek the students' points of view in order to understand students' present conceptions for use in subsequent lessons.
Students primarily work alone.	Assessment of student learning is interwoven with teaching and occurs through teacher observations of students at work and through students' exhibitions and portfolios.

Source: de Esteban and Penrod (2000).

CHAPTER 3

THE ABUSE AND MISUSE OF THE STANDARDS MOVEMENT

The American public school system has always had a method of measuring whether students were meeting standards. Traditionally it has been success upon graduation in the work force. To what extent is the system turning out productive citizens who can earn an income and contribute to society? The progressive education movement had at its heart the desire to expand democracy, sympathy for the immigrant poor, and an attempt to counterbalance unbridled wealth of the bosses with an educated populace, driving against municipal corruption; all very noble indeed. With the rise of capitalism at the turn of the century, America became more affluent while reaping the rewards of a democratic educational system. It also brought with it the abuse of immigrants, who struggled in the manual labor workforce to build transportation and infrastructure systems in burgeoning cities across the nation. A true democratic education system would balance out the social status of the citizenry and ensure that caste or birthplace would not prevent a person from realizing his or her potential, as was the case in so many other countries. Somewhere along the way in the mid-twentieth century, this means of measuring educational achievement began to lose its effectiveness, and the American system needed some other way to show how our students were doing. History chose science as the backdrop for historical change. This was brought to the attention of the American public

The Unfinished Quest: The Plight of Progressive Science Education in the Age of Standards
pp. 45–51

like a bolt of lightening and was to rock the complacent school system to its foundation.

On October 4, 1957, the Soviets trumped America's space effort by launching Sputnik I, the world's first artificial satellite. The result was the birth of the space race. The space race grew out of the Cold War between the United States and the Soviet Union. For 50 years the two superpowers struggled for global supremacy. Space was a crucial arena for this battle.

An angry Harry Truman blamed the persecution of American scientists by Joseph McCarthy in the 1950s as the reason the American scientists fell behind in the development of rockets and satellites. Immediately after its launch, a humiliated U.S. Defense Department reacted by approving funds for an American version of a satellite; resulting in the launching of Explorer I on January 31, 1958. Sputnik also spurred the creation of the National Aeronautics and Space Administration in October, 1958.

A second result of the Space Race was a formulation of a committee to address prevalent questions and problems in science education. In September of 1959, 35 educators, scientists, and scholars gathered at a conference at Woods Hole on Cape Cod to discuss how science education might be improved in America's schools. The 10-day meeting was called by the National Academy of Sciences which had been examining through its education committee, the long-range problem of improving access to scientific knowledge in America. The intention was to examine the fundamental processes involved in imparting to students a sense of the methods and foundations of science (Bruner, 1960). This was the first time psychologists had been brought together with scientists to discuss problems involved with teaching various disciplines. The major topic of discussion was how children learn science.

Despite all of this catch up work though, the U.S. educational system began to suffer a series of attacks from politicians and others angry with the notion that American schools must be inferior to Soviet schools to result in this defeat. So began a long series of events that culminated in the modern standards movement.

In the early 1980s, a new movement in education was born of this crisis. It has been dubbed "The Excellence Reform Movement" backed by Ronald Reagan, and was triggered by *A Nation at Risk*; a report drawn up by the U.S. Department of Education. Since the publication of *A Nation at Risk*, a number of reform efforts in science education have resulted in the improvement in the average scores in science. This report focused on global competition in the workforce by American citizens, so the agenda was economic. This worked on the assumption of course, that the U.S. school system had failed, and that the only answer was sweeping educational reform. The report compared the state of American education with being at war (M. Berube, 1994). President George Bush declared an agenda for reform titled *Goals 2000* with six major objectives whereby the United States would be the first in international comparisons. President Bill Clinton continued the agenda, adding new goals such as graduation rates and literacy (U.S. Department of Education, 1994).

Among the findings of *A Nation at Risk*, were the following indicators of risk:

- Twenty-three million Americans are illiterate.
- Achievement of high school students are lower than when Sputnik was launched.
- Over one half of gifted students do not work up to their ability.
- SAT scores have declined from 1963 to 1980.
- Too many teachers are drawn from the bottom quarter of graduating high school and college students.
- Half of the teachers in subjects like math, science, and English are not qualified to teach their subjects.
- Math skills are lower.
- The curricula have been watered-down homogenized, diluted, with no central purpose.
- Too many "methods" courses are taught in teacher education colleges, not enough content ("A Nation at Risk," 1983).

Critics of progressive education are quick to defend this tome, blame of the state of U.S. education on those who claim to be either progressive or constructivist in their approach, citing lack of discipline, lower content knowledge, "out of control" classrooms, and not enough of the classics. In his book *Getting It Wrong From The Beginning*, Kieran Egan (2002) argues that the leading influential minds of progressive education, from Piaget and Dewey to Gardner, are all misguided in stating that the psychology of children and theories of child development are important components of successful educational strategies. Egan claimed that these ideas of human psychology as an important tool in our understanding of educational practices was flawed. Egan states that we should abandon any attempts to try to learn how children prefer to learn, a "natural" best way, and that rote learning doesn't deserve the negative connotation it receives (pp. 43-48).

Egan (2002) proceeds to criticize even these following educational practices:

- Whole language
- Children becoming personally and actively involved in their education.

- Children having a natural inclination to explore and manipulate their environments.
- play as valuable experience
- It is helpful to find concrete ways of relating new knowledge to what they already know.
- Children are more likely to understand and remember information when they discover it themselves.
- Learning takes place following sensory stimulation from people, things, and activity in the environment (p. 43) Indeed, it is hard to find anything wrong with these educational practices, delivered by a competent teacher.

Egan (2002) particularly focuses on Herbert Spencer, the Victorian philosopher and sociologist, whose ideas of "social evolution" mirrored Darwin's ideas of biological evolution. His ideas that children are naturally good learners went against the Victorian notions at the time that humans and children in particular are evil by nature and need to be "trained" out of it. He posited that if we could discover how to harness the natural curiosity of children, learning would be made easier. Egan considers Spencer an influential figure to Dewey, since most of Dewey's works mirror Spencer's ideas. What is curious is Egan's rabid defense of "rote learning," which both Spencer and Dewey denounced.

In a section of his book titled "The Joy of Rote Learning," Egan (2002) goes on to state that learning by "rote" is merely another way of saying, "learning by heart," which I believe has two vastly different meanings (p. 67). Indeed, I believe they mean just the opposite. To learn by "heart," means that the learner has completely mastered a concept, facts, concepts and meaning to where memorization is no longer necessary when asked later to retrieve the information. "Rote" means that the student memorized something and can regurgitate the facts without necessarily knowing the meaning behind them. I had this experience myself while studying doctoral level statistics. Students in my college gathered in study groups led by tutors, since we were to be tested in this topic in order to proceed to the next level and begin work on our dissertations. My first tutor relied on having us students memorize concepts and applications, and needless to say, I was beyond myself with frustration. A number of us switched to another tutor, a psychology grad student, who did not rely on pure memorization, but laid out concept maps for us to see how everything was related. This was a brilliant breakthrough for me, because now I could completely understand the conceptual topics at hand (*why* certain statistical methods were used for certain reasons) and did not have to rely anymore on rote memorization. Many times in education, students fail to learn not because of lack of intelligence, but because the teacher fails to connect the learning for the student in meaningful ways (or to show the student how to make the connections themselves). How things relate to each other is crucial for higher order learning to occur. This does not mean that I did not have to memorize certain things, but it all made sense when I saw the big picture, and I did not have to rely on memorization as the sole method for content retrieval later during the test.

Needless to say, there is much criticism of progressive and constructivist education, most of it stemming from the very practices and behaviors constructivist teachers employ. The following is a short synopsis of the differences in teaching behaviors between the two camps. I will call them constructivists to represent the progressives, and traditional to represent the standard pedagogy.

Constructivists

- child-centered (instead of teacher-centered)
- discovery learning (teacher knows high content levels, but lets students' discover answers)
- scaffolding (teachers act as support, instead of main star in the classroom)
- schema (students' ideas of concepts, that they bring to the classroom)
- alternative assessments (other than pencil and paper tests)
- theory of multiple intelligences
- hands-on learning
- gender related learning preferences
- problems-based (process)
- cooperative learning

Traditional

- teacher centered
- teacher telling
- teacher is main star, students listen
- traditional assessments (standardized tests)
- intelligence theories dismissed
- some hands-on learning

- no thought given to gender issues
- product based
- some cooperative learning

Of course, everything has degrees; so too does each philosophy. A good progressive teacher should use well-planned lecture in many instances, just as a good traditional teacher should use cooperative learning on occasion. But it has been the author's experience after several years in the classroom, each style of teaching stays closely aligned to the stereotypical list of behaviors such as the one above. One area with which both sides agree is that teachers should possess great content knowledge in order to teach any subject.

Indeed, it would be a foolish person who argues that we need teachers with less content knowledge in our classrooms, yet this is the one major criticism of constructivist teachers. Maybe the problem with the lack of teachers with content knowledge points more truthfully to the fact that mostly women teach elementary school, yet most women are very uncomfortable teaching science, regardless of pedagogy preference. Maybe if teachers were paid more and thus given more respect, scientists in the field could be lured away from their lucrative careers to teach in the schools. There are obviously serious problems with the lack of content-knowledge issue, and it is easy to cast the blame where it does not belong. True progressive education values highly educated teachers who are experts in their fields. But progressive pedagogy involves constructive practices, which combine content knowledge with higher order learning. Constructivist teachers can be compared to jazz musicians. Most people improperly believe that jazz musicians "don't know their music" because they do not seem to follow normal musical rules. Indeed, jazz musicians not only know their music, but know it at higher levels than most other musicians, and can break from the rules and soar to sublime levels of musical brilliance. One cannot rise above the rules without knowing the rules by heart. It is those most uncomfortable with the rules that stick most closely to them.

Unfortunately, it is here where progressivism and standards collide, since the assessment of the standards has become more important than the learning itself.

As was mentioned in chapter 2, Benjamin Bloom (1956) devised a way to measure the level at which a teacher was teaching. From knowledge to evaluation, teachers could chose a level of Bloom's at which to teach a lesson, use the corresponding verbs provided by Bloom when writing daily objectives to ensure the level was being taught, and then assess that objective. For example, if Ms. Smith chooses to teach a knowledge-based lesson on plant cells, she could write a daily objective for her lesson as:

"the student will be able to list and label the parts of the plant cell on the diagram." Then Ms. Smith could collect the work and assess her objective by grading the diagram and in so doing, would be able to tell who mastered this assignment. It is extremely easy to assess the lower levels of Bloom's (1956) taxonomy, namely knowledge and comprehension. The higher one goes up Bloom's levels, the more difficult it becomes to assess, since the assessments become more subjective in nature. When Ms. Smith grades an essay for example, which measures analysis and synthesis, it becomes less and less clear how to properly grade, since arguments can be made on each side for each mark given. Indeed, meaning becomes very important at this level; the student can actually solve a problem or answer a question in a *different* way than the teacher expected, but it could be a better answer or better method of problem solving than the teacher had in mind. If one is grading purely objectively, there is no room for "different" answers, which in themselves might be quite brilliant. I have experienced this occurrence many times in academia, where my answer did not quite suit the teacher's, and even though quite imaginative, given a failing mark. As a teacher, I have had students who delivered an answer that was better than my solution, and thankfully was not the kind of teacher to penalize for creative thinking. Although this is the more desirable method of assessment, it is also the more difficult. (It should be said here, that it is assumed that if a child masters the higher levels of Bloom's, they also have mastered the lower levels; the same cannot be said in reverse.)

So how exactly do the standards collide with progressive education? The answer is in how the standards are assessed.

CHAPTER 4

STANDARDS AND SCIENCE CONTENT COMPREHENSION

The Consequences of High-Stakes Testing

American schools are driven by money. The better school districts are those that have the most money. Better housing means more taxes which means more money for the schools. The federal government is supposed to make up for this inequity, but all one has to do is to visit an inner-city school in the poorer part of any town or city, to see that they do not.

The 1983 report, *A Nation at Risk,* initiated public awareness of standards. Indeed, raising standards was named as one of the five major recommendations in the report. In 1989, George Bush Sr. held a National Education Summit of State Governors to establish six broad goals to address the issues raised by *A Nation at Risk*. This summit emphasized development of standards and student performance. There were a series of governors summits, culminating in the 2000 No Child Left Behind (NCLB) act, which was a revised version of the 1965 Federal Elementary and Secondary Education Act, signed into law by George W. Bush. The NCLB act tied meeting state educational standards to federal funding. This changed the federal government's role in K-8 education by asking each school to describe their success in terms of what each student

The Unfinished Quest: The Plight of Progressive Science Education in the Age of Standards
pp. 53–69

accomplishes. These accomplishments have to be measured somehow, so each state set out to develop it's own set of standardized tests that would accomplish this goal.

In Virginia, for example, the Standards of Learning (SOL) tests are given at the end of the year in Grades 3, 5, 8, and at the end of each subject area taken in high school. In Grade 3 for instance, students are tested in English, math, history, social studies, and science. These tests are held in May and take up a significant amount of instructional time, both in their preparation and their implementation. Grade 5 students are tested in English: reading, literature, research, writing, math, history, social studies, science, and computer technology. Grade 8 students are tested in the same subjects as the fifth graders. In high school, each separate subject gets it's own test, including English: reading, literature, writing, Algebra I, Algebra II, geometry, Earth science, biology, chemistry, World History I, World History II, world geography, and U.S. history (Virginia Department of Education [VDOE], n.d.).

So far, so good, no one can argue with the premise of *A Nation at Risk*, or NCLB, both noble undertakings. And no one would deny that assessments need to be given for reliable measurement of student progress. But let's examine the assessment instruments themselves and how they actually measure achievement.

I could not have taught middle school science without my copy of the Virginia SOL in hand. They are written so that teachers can have a full grasp of the scope and sequence of the topics and concepts in their subjects that they are to teach for the entire year. The Virginia SOLs for seventh grade science were my invaluable guide as I prepared my lessons for the year, and I would have been lost without them. Here are the Virginia SOL for seventh grade science (life science)

LIFE SCIENCE

The Life Science standards emphasize a more complex understanding of change, cycles, patterns, and relationships in the living world. Students build on basic principles related to these concepts by exploring the cellular organization and the classification of organisms; the dynamic relationships among organisms, populations, communities, and ecosystems; and change as a result of the transmission of genetic information from generation to generation. Inquiry skills at this level include organization and mathematical analysis of data, manipulation of variables in experiments, and identification of sources of experimental error.

The Life Science standards continue to focus on student growth in understanding the nature of science. This scientific view defines the idea that explanations of nature are developed and tested using observation, experimentation, models, evidence, and systematic processes. The nature of science includes the concepts that scientific explanations are based on logical thinking; are subject to rules of evidence; are consistent with observational, inferential, and experimental evidence; are open to rational critique; and are subject to refinement and change with the addition of new scientific evidence. The nature of science includes the concept that science can provide explanations about nature, can predict potential consequences of actions, but cannot be used to answer all questions.

LS.1: The student will plan and conduct investigations in which

(a) data are organized into tables showing repeated trials and means;
(b) variables are defined;
(c) metric units (SI—International System of Units) are used;
(d) models are constructed to illustrate and explain phenomena;
(e) sources of experimental error are identified;
(f) dependent variables, independent variables, and constants are identified;
(g) variables are controlled to test hypotheses, and trials are repeated;
(h) continuous line graphs are constructed, interpreted, and used to make predictions;
(i) interpretations from a set of data are evaluated and defended; and
(j) an understanding of the nature of science is developed and reinforced.

LS.2: The student will investigate and understand that all living things are composed of cells. Key concepts include

(a) cell structure and organelles (cell membrane, cell wall, cytoplasm, vacuole, mitochondrion, endoplasmic reticulum, nucleus, and chloroplast);
(b) similarities and differences between plant and animal cells;
(c) development of cell theory; and
(d) cell division (mitosis and meiosis).

LS.3: The student will investigate and understand that living things show patterns of cellular organization. Key concepts include

(a) cells, tissues, organs, and systems; and
(b) life functions and processes of cells, tissues, organs, and systems (respiration, removal of wastes, growth, reproduction, digestion, and cellular transport).

LS.4: The student will investigate and understand that the basic needs of organisms must be met in order to carry out life processes. Key concepts include

(a) plant needs (light, water, gases, and nutrients);
(b) animal needs (food, water, gases, shelter, space); and
(c) factors that influence life processes.

LS.5: The student will investigate and understand how organisms can be classified. Key concepts include

(a) the distinguishing characteristics of kingdoms of organisms;
(b) the distinguishing characteristics of major animal and plant phyla; and
(c) the characteristics of the species.

LS.6: The student will investigate and understand the basic physical and chemical processes of photosynthesis and its importance to plant and animal life. Key concepts include

(a) energy transfer between sunlight and chlorophyll;
(b) transformation of water and carbon dioxide into sugar and oxygen; and
(c) photosynthesis as the foundation of virtually all food webs.

LS.7: The student will investigate and understand that organisms within an ecosystem are dependent on one another and on nonliving components of the environment. Key concepts include

(a) the carbon, water, and nitrogen cycles;

(b) interactions resulting in a flow of energy and matter throughout the system;
(c) complex relationships within terrestrial, freshwater, and marine ecosystems; and
(d) energy flow in food webs and energy pyramids.

LS.8: The student will investigate and understand that interactions exist among members of a population. Key concepts include

(a) competition, cooperation, social hierarchy, territorial imperative; and
(b) influence of behavior on a population.

LS.9: The student will investigate and understand interactions among populations in a biological community. Key concepts include

(a) the relationships among producers, consumers, and decomposers in food webs;
(b) the relationship between predators and prey;
(c) competition and cooperation;
(d) symbiotic relationships; and
(e) niches.

LS.10: The student will investigate and understand how organisms adapt to biotic and abiotic factors in an ecosystem. Key concepts include

(a) differences between ecosystems and biomes;
(b) characteristics of land, marine, and freshwater ecosystems; and
(c) adaptations that enable organisms to survive within a specific ecosystem.

LS.11: The student will investigate and understand that ecosystems, communities, populations, and organisms are dynamic and change over time (daily, seasonal, and long term). Key concepts include

(a) phototropism, hibernation, and dormancy;
(b) factors that increase or decrease population size; and

(c) eutrophication, climate changes, and catastrophic disturbances.

LS.12: The student will investigate and understand the relationships between ecosystem dynamics and human activity. Key concepts include

(a) food production and harvest;
(b) change in habitat size, quality, or structure;
(c) change in species competition;
(d) population disturbances and factors that threaten or enhance species survival; and
(e) environmental issues (water supply, air quality, energy production, and waste management).

LS.13: The student will investigate and understand that organisms reproduce and transmit genetic information to new generations. Key concepts include

(a) the role of DNA;
(b) the function of genes and chromosomes;
(c) genotypes and phenotypes;
(d) factors affecting the expression of traits;
(e) characteristics that can and cannot be inherited;
(f) genetic engineering and its applications; and
(g) historical contributions and significance of discoveries related to genetics.

LS.14: The student will investigate and understand that organisms change over time. Key concepts include

(a) the relationships of mutation, adaptation, natural selection, and extinction;
(b) evidence of evolution of different species in the fossil record; and
(c) how environmental influences, as well as genetic variation, can lead to diversity of organisms (VDOE, n.d.).

These are the standards that seventh grade life science teachers in the state of Virginia must adhere to meet Virginia curriculum guidelines. While they address several topics and concepts, they are broad enough so that the teacher can construct his/her teaching in many different ways, so long as she meets the standards. Notice that they are written as "the student will investigate and understand." While the students will eventually be able to understand the concepts through investigation, it is up to the teacher to determine just how the students will understand. This is the link between the global SOL standard, and the more specific daily objective the teacher writes into her daily lesson plans. "Understand" is not a specific enough verb to use when writing daily lesson plans. How will the student understand exactly? How will you *know* the students understand? The daily objectives must be specific, and they must be *measurable*. Seventh grade life science teacher Ms. Thompson could write the following daily objectives to match LS 1, bullet a and b for example:

> Daily objective #1: The student will be able to construct a data table.
> Daily objective #2: The student will be able to define variable, including dependent, independent, confounding, and will be able to write 10 problem statements using independent and dependent variables.

Now, daily objective #1 might cover a single lesson plan that covers an entire day or session. Or he might combine objectives #1 and #2 into a single lesson plan that takes more than 1 day to complete. Ms. Thompson could also write her daily objectives using verbs from any of the six levels of Bloom's (1956) taxonomy, thereby ensuring that certain cognitive levels will be taught. If the verbs "list," "define," or "match" are used, you can guarantee she is teaching at the knowledge level, the first or lowest level of Bloom's taxonomy. Ms. Thompson could also use verbs such as "plan," "write," or "discriminate," thus ensuring that he is teaching at the evaluation level, Bloom's highest cognitive level. This is the professional freedom teachers have historically had over their teaching methods. However, more and more lately, teachers are forced to revert to "teaching to the test" in order to prepare their students for the SOL test, given in May every year.

"Teaching to the test" usually refers to the practice of teaching students in a specific way so that they will be practiced enough in test taking skills so that they will be able to pass the test. It involves test-taking strategies, how to take multiple choice tests, how to properly read test questions, how to pace yourself during test time, and a host of other strategies that have little or nothing to do with the subject matter being taught. If there are any who doubt that scores can rise by studying how to take a multiple choice test, all one has to do is to inquire about Stanley Kaplan test prep and admissions Web site, to learn all about how your child can raise his or her score on the SAT or a multitude of other tests by taking their test-taking classes, to the tune of 700 or $800 (Kaplan Test Prep, 2005). The point I am making is that there are ways to instruct students that will better prepare them for taking a test, and there is nothing wrong with this as a concept or practice. Where the problem arises in the public school system, is when teachers abandon "best-practices" and teaching higher-order thinking skills to spend a month preparing their students to take a test in order to meet and pass the state standards test.

Another problem is that the tests cover material from as far as two or three years back. For the eighth grade physical science test for the Virginia SOLs, part of the test covers topics covered in sixth grade science, part covers seventh grade science, and part covers the eighth grade physical science. So not only do the teachers have to prepare the students to pass by reviewing material covered that year, but she has to go back several years to refresh the students on topics already covered, and already passed in school. This is an enormous waste of classroom time and it detracts from new topics and concepts the students could be learning instead.

Do Rising Test Scores Mean Higher Comprehension?

Of course, the assumption is that if scores are going up as they are in Virginia (131), that the students are learning more, comprehending the material, and becoming better prepared for the economic workforce. I recently did a study called *Are Standards Preventing Good Teaching* (C. Berube, 2004) where I discovered some disturbing things about the yearly standards tests.

My focus for this study was to see if indeed students were learning more since the standardized test scores were rising. In this case, the standardized test I chose to study was the Virginia SOL test for eighth grade physical science. My concerns going into this study were numerous. Did teaching style contribute to test scores? Does passing the test mean higher comprehension? Are the SOLs being misused as high-stakes tests? My hypotheses were more specific: I believed that constructivist teachers would have students who scores higher on the SOL test, since research indicates better delayed post-test retention of material of students taught by constructivist teachers (Bay & Others, 1992).

The second measurement instrument I measured was one I created, called the Higher Order Skills (HOS) test. This test was a simple instrument. It was the same SOL test the students were taking, except that after each multiple choice question, the students were asked to explain their answers. Items were identical to the SOL items, except that in addition to being required to circle the correct answer as in the regular multiple-choice format, students also were asked to explain their answer. At this point, comparison was made between the scores from the SOLs and the HOS measurement. The HOS instrument measured comprehension. Comprehension, according to Bloom, refers to the type of understanding such that the individual knows what is being communicated and can make use of the material or apply the idea. Comprehension enables translation such that the material in its original communication is preserved although the form of the communication has been altered, as in transferring multiple-choice answers into short-essay format. Comprehension is also the ability to understand non-literal statements and to interpret and extrapolate (Bloom, 1956). In short, the HOS instrument asked the students to explain and extrapolate their multiple-choice answer.

The sample for this study was taken from urban middle schools in a Southeastern urban school system, namely Norfolk, Virginia. The sample included Caucasian and minority middle school science students. The study was conducted in 13 intact classrooms, three employing traditional instructional methods, five employing progressive/constructivist instructional methods and five employing mixtures of both, designated as mixed. Type of classroom was identified through surveys where teachers stated which type of instruction is employed in their classroom. Every eighth grade science teacher in the district was asked to complete a copy of The Revised Constructivist Learning Environmental Survey (Taylor, Fraser, & White, 1994), which measures teacher perception of constructivist attributes in the learning environment. This instrument is designed to measure the constructivist approaches used in teaching science. The results of this survey provide insights into classroom environments and pedagogical basis of instruction. It is a 30 question, 5-point rubric or Likert questionnaire which identifies teacher perception of the presence of characteristics of constructivism on five subscales, with six questions each: personal relevance, scientific uncertainty, critical voice, shared control, and student negotiation. The composite scores were used to determine a "low" (score of 30) or "high" (score of 150) degree of constructivism in the classroom environment.

Upon return of the surveys, the researcher observed each classroom and scored a copy of CLES (Constructivist Learning Environment Survey) to match up teacher perceptions to researcher perceptions. The 13 classrooms were ranked from most constructivist to most traditional.

The independent variables are:

1. Teacher type (with three levels: constructivist, mixed, and traditional).
2. Ethnicity (with two levels—Caucasian vs. minority).
3. Gender (with two levels—male vs. female).

The dependent variables are:

1. SOL scores
2. HOS scores

The primary analysis was MANOVA (Multivariate Analysis of Variance) through which differences in SOL and HOS associated with the different methods and with race and gender were determined.

There were limitations brought on by the self-scoring instrument. Perceptions of the constructivist learning environment are self-reported measures, and it can never be certain if the teachers' responses are true reflections of their attitudes, perceptions or behavior. Teachers may respond in socially desirable ways and may not be good observers of their own behaviors in the classroom, and, in turn, may think they are teaching a certain way but are not. This was minimized by clear questionnaire definitions as to specifically what are the two types of instructional methods and the characteristics of each (objectives), and also by my observation of the classrooms after the self-assessment.

The first step in implementing this study was that middle-school principals were contacted for consent. Principals were sent consent letters for them to sign and return. Upon obtaining consent from five of the eight principals, The revised CLES was delivered to eighth grade physical science teachers in those schools, along with teacher consent forms to sign and return. The population consisted of 13 teachers.

The CLES survey scores were used by myself to determine which classrooms were traditional and which were constructivist. From the 5 middle schools, 13 teachers were available and sent in self-scored CLES surveys. I went into these classrooms and observed for teaching styles and completed extensive notes for each teacher.

My observations were conducted over a period of 4 weeks. I observed each classroom for a period of approximately 3 hours. Extensive notes were taken and analyzed for evidence of constructivist practices and classroom setting.

Upon observation of the classes, I distributed copies of the HOS test to each teacher with directions to administer a week after the SOL test. SOL scores were obtained by contacting each school for a viewing of the printout of the results. The answers of the SOL tests and the comprehension measurement were compared to determine if comprehension is greater with those students taught in the constructivist classrooms versus the traditional classrooms.

What I discovered was unsettling. My hypothesis was wrong concerning constructivist teachers having students with the higher SOL and HOS scores. As it turned out, the more mixed and traditional teachers had the highest scores. For this to be the case, one of two things was occurring: either traditional teachers really do have skills that result with students having higher content knowledge, which goes against most of the literature, or that during stressful high-stakes test time, the teachers with more traditional methods employing fact-based drill and lecture, were better suited to preparing their students for multiple-choice standardized tests.

But the most startling finding was something that I did not expect. I decided to concentrate on those students who *only* passed the SOL test. Then I matched those same passing students' scores to their HOS scores. Much to my dismay, almost 75% of those students who passed the SOL, failed the HOS exam, which was the same exam remember, only they had to explain their answers. Did this mean that the students had memorized certain possible questions ahead of time? Had their teachers engaged in "teaching to the test" in order for them to pass? Were the more constructivist teachers buckling under the pressure during test preparation time and abandoning their teaching practices for more knowledge based rote drill test preparation? Many questions were raised in my mind as to how so many students could pass the much touted SOL, but fail to explain their answers?

One of the explanations might be that when standardized tests are employed as high-stakes tests, instruction suffers. Studies by Lewis (2003) and Schroeder (2003) showed that tests often effect instruction in ways that directly contradict the intent of state educational reform policies to raise standards. When administrators put pressure on teachers, those teachers may resort to a teaching mentality that is conducive to passing a test, instead of higher-order learning and problem solving. And if those teachers and administrators could be transferred or lose their jobs altogether, the implications become obvious.

Inductive/Conceptual Science Learning

In a recently published article in the Virginian-Pilot, June 3, 2007, titled "Why Are So Many Students Acing Some SOL Tests?", the author Amy Jeter describes a situation in Virginia public schools where in some schools, as much as half of the students were not only passing the tests, but "acing" them, or scoring perfect scores. The subject? History. According to Jeter, 56% of third graders in a particular school in Chesapeake, Virginia, aced the test. Across the entire state, one student in five did the same. But the other content areas did not offer such easy As, including science, which had the lowest "ace" rate in the state. The nature of science which I have been stating throughout this book, is clearly more discovery oriented in nature than history, and less able to be correctly measured by a multiple choice test.

According to Jeter (2007), students have not always sailed through the history tests. When they were first given, scores were so low, that the Virginia Board of Education lowered the number of required questions needed to pass the test. So clearly we have a situation where students over the years were "test practiced" and taught how to more effectively take a multiple choice test. In science, the number of students passing them hasn't changed, but the numbers scoring perfect scores has remained low, about 5.4% for third graders, and 2.3% for fifth graders. Science learning by it's very nature is inductive, and again, not measured properly by the standardized tests. "Science tends to require problem-solving rather than fact memorization" (Jeter, 2007). Ashanda Bickham, a teacher at Norfolk's Chesterfield Academy of Math, Science and Technology, puts it this way, "They have to know the concept, and they have to be able to apply it ... it's a higher level of thinking." So conceptual learning is not being measured correctly by the Virginia SOLs.

There is another, more insidious problem plaguing science classrooms across the country, and that is the fact that most elementary and many middle school science teachers to a great extent are intimidated and afraid of science, and don't have the proper blocks of time needed to teach in-depth labs and experiments. Elementary teachers tend to be generalists, and science tends to be the least favored subject among them. As a result, the cycle continues—poorly trained elementary science teachers, doing a less than stellar job teaching science to elementary children, who therefore enter middle and high school with less than stellar credentials and backgrounds for higher level science courses.

Possibly the most valuable skill a scientist can have is this inductive thinking pattern. In his new book *Einstein; His Life and Universe,* Walter Isaacson (2007) brilliantly describes this very trait in the most famous and important scientist of our time, Albert Einstein. Einstein developed his theories for relativity as a child by visualizing himself riding a beam of light. His theories were developed as complete whole insights, and details were worked out later. Indeed, Einstein's strength was his ability to visualize problems. His famous "Gendanken" thought experiment of the twins paradox that describes how traveling near the speed of light can slow time down so a person will not age, came about as a visualization. His visualizations were amazing because he could "see" whole complete theories in his head. He once mused that since the mathematicians had gotten hold of his theory of relativity, he no longer understood it! I believe we are teaching science backwards, teaching meaningless disjointed facts first, then finding theories to fit them, instead of the much more fun and interesting way to learn science, the inductive methods which Einstein employed. The problem with this is that we do not know how to assess this type of creativity, we only know how to measure the lowest levels of learning, the knowledge level.

In 2000 I wrote and published an article dealing with this very topic, titled "A Conceptual Model of Middle School Science Instruction" where I argue that middle school science is being taught backwards, with rote facts being learned first, while concepts connecting the facts are being neglected. I firmly believe that science is more alive, interesting, and creative when facts are memorized as a result of inductive learning, instead of just memorizing to memorize. Einstein would agree with me.

High-Stakes Tests?

Were these tests used to measure the standards movement ever meant to be such high-stakes in nature? Diane Ravitch (1996) says no. Ravitch is a former assistant secretary of education for research under George Bush, currently a research scholar at New York University. In her piece "The Case for National Standards and Assessments," she makes a noble case for standards and puts the argument into perspective. As I mentioned earlier, standards are good things. We need to have them. But we have lost sight of what they are supposed to be about.

Ravitch states that the following claims have been made by supporters of standards:

- Standards can improve achievement by clearly defining what is to be taught and what kind of performance is expected.
- Standards are necessary for equality of opportunity.
- National standards provide a valuable coordinating function (bringing together the pieces of the educational system to operate coherently).
- Standards and assessments provide consumer protection by supplying accurate information to students and parents.
- Standards and assessments serve as an important signaling device to students, parents, teachers, employers, and colleges (Ravitch, 1996).

I believe this is the true nature of standards. Nowhere here does Ravitch (1996) mention that assessments should be of such a high-stakes nature, such as they have evolved into. When tests become stressful, they lose their purpose as assessments, to gauge where students are and how best to help them, instead they become the end in themselves, and the passing of them becomes more important than the learning.

In Noll's (2005) excellent book, *Taking Sides,* he pits against each other those in favor of high-stakes tests as assessments to improve learning, and those who say such tests do not improve learning. Among those arguing against high-stakes tests, in the section titled "The Courage to Be Constructivist," are Martin G. Brooks and Jacqueling Grennon Brooks. They argue that constructivism powerfully informs educational practice by describing the central role that learner's ever-transforming mental schemes play in their cognitive growth. They, like myself, argue for standards and accountability, but make an important point by stating,

> We are not suggesting that educators should not be held accountable for their students' learning. We believe they should. Unfortunately, we are not holding our profession accountable for learning, only for achievement on high-stakes tests. As we have learned from years of National Assessment of Educational Progress research, equating lasting student learning with test results is folly. (p. 162)

This speaks directly to my study, where I learned first-hand that students could pass the SOL test with flying colors, but not understand what they learned.

Here are some findings from the research presented by Brooks and Brooks (1993):

- Test scores are generally low on the first assessment relating to new standards. Virginia is an extreme example of this phenomenon: More than 95% of the schools failed the state's first test. In New York, more than 50% of the state's 4th graders were deemed at risk of not graduating in 2007 after taking that state's new English language arts test in 1999.
- Failure, or the fear of failure, breeds success on subsequent tests. After the first administration of most state assessments, schools' scores rise because educators align curriculum closely with the assessments, and they focus classroom instruction directly on test-taking strategies.
- To increase the percentages of students passing the state assessments, and to keep schools off the states' lists of failing schools, local district spending on student remediation, student test-taking skills, and faculty preparation for the new assessments increases.

Despite rising test scores in subsequent years, there is little or no evidence of increased student learning. A recent study by Kentucky's Office of Educational Accountability suggests that test-score gains in that state are a function of students' increasing skills as test takers rather than evidence of increased learning" (Hambleton et al., 1995).

Is the rise of standardized testing as a means of measurement for state standards, simply an effort on the part of the federal government to "standardize" or control student learning, or is it a valid way to ensure that all of our students are learning at the highest levels? Constructivism tells us that while we can control what we teach, we cannot control what the students learn (Noll, 2005, p. 162). All one has to do to see this in action is to face a classroom of 25 students and deliver the same lesson and see what happens. There is no way the government can control what students learn. While they can ensure all students are learning the same topics and meeting the same federal guidelines for curriculum, one must remember that each student will make sense of the curriculum in his or her own way, with the teacher's help.

There is also the danger of special interest groups controlling public school curriculum in order to match it up to the testing. Which theory of the origins of life should be taught? Should they all be included in the standardized tests? What if some of the topics covered in the standardized tests offend certain public school constituencies? Can the U.S. government regulate moral curriculum and include it in the standardized testing?

Assessing the Standards

Content standards are necessary if we want to have a roadmap towards student improvement. They are the starting point for the educational process, just as the finish line at the Boston Marathon is the point where every marathon runner begins his or her training, with that in mind. According to Diane Ravitch (1996), without standards, you leave the decision as to what is taught in our nation's schools to textbook companies, testmakers and interest groups.

In 1994, Congress passed a law to begin process of creating national content and performance standards, but *assessments* were left up to the states. In science and math, all states need the same standards, so international standards have been developed, but this is important for people in the sciences and math fields who move from state to state. The concern here, is that if assessments are left up to the states, who is policing the states to ensure they are being enforced correctly, and that all states are adhering to the same guidelines?

Different states have different water-mark levels that determine proficiency. But this is a problem. According to McCombs, Kirby, Barney, Darilek, and Magee (2004) and McCombs and Carroll (2005):

> Equally dubious is the meaning of "proficiency," the definition of which varies from state to state. To validate the state test results, therefore, students must also take a national test, called the National Assessment of Educational Progress (NAEP). Known as the Nation's Report Card, the NAEP is the only common criterion by which to compare student achievement across the states. Statewide proficiency rates on the NAEP are often much lower than on the state tests. (p. 10)

For example, In Virginia, the state results in third grade revealed that 71% of students were proficient in English; while the NAEP scores for fourth grade reading proficiency were a mere 37%. In math, the percentage was 87%, the NAEP percentage was 40%. In eighth grade, the Virginia SOL test results showed that 72% of students were proficient in English and 78% in math; the NAEP scores were 35 and 33% respectively. Not only in Virginia, but in several other states as well. In Mississippi, 89% of fourth graders were proficient in state reading tests, while the NAEP recorded only 18%. In New York, the state test had 70% proficient in reading, while the NAEP test showed only 33% proficient in reading. This disparity of scores between state and NAEP tests occurred in 20 states ("Education Law Gets First," 2005).

These findings have provoked some educators to recommend a federal takeover of standards and testing. In a recent *New York Times* piece, Diane Ravitch (2005) criticized the "50 states, 50 standards, 50 tests" as a failed strategy because the "evidence is growing that this approach has not improved student achievement." She called for a national takeover of standards and testing. National standards and national testing imply a national curriculum that may be resisted by states and localities. Ravitch proposes one-third national, one-third state, and one-third local (M. Berube, 2005; Ravitch, 2005). This would leave room for not only national comparisons of standards, but also for individual municipalities to include items on the exam that would be specific to that area and would better deal with cultural, racial and gender issues. This may be the answer for the future of standardized testing.

The Real Purpose of Assessment

The purpose of any assessment is to *assist* student learning, not to be used punitively or in a strictly summative fashion. It is meant to identify student's strengths and weaknesses (diagnostic), measure the effectiveness of particular instructional strategies, measure and improve the effectiveness of curriculum programs, measure and improve teaching effectiveness, and to provide data that assists in decision making (prescriptive). Progressive education and constructivism in particular, is concerned with the *process* of learning, in addition to the product.

Have you ever had this experience in school? You work out a math problem, and you come up with the correct answer, but you did not arrive at the solution *the teacher's way*, so even though the answer is correct, you lose points or get it marked wrong altogether. This style of thinking is closed-minded indeed, and thwarts many an imaginative student from finding creative solutions to problems, yet this goes on in almost every school in every city across America every single day. This is a closed-ended approach to teaching, and no great innovation or idea ever got it's start this way. But this is exactly how the standardized tests are evaluated.

Once when I was teaching a college class about structuring their lessons around their objectives to meet their obligations with the standardized tests, but to also teach at high levels, one student answered that "in the real world," they could be fired for not teaching to the test so their students could pass it. I replied that "in the real world," there was a situation that occurred on April 13, 1970, where the Apollo 13 spacecraft suffered an explosion and started to lose oxygen, threatening the lives of the crew. NASA in Houston ordered a group of scientists to an emergency meeting on the ground, and dumped onto a table all of the available equipment the crew would have onboard. They were then asked to "solve this problem." The equipment was imperfect. The crew had a limited supply of materials, and not the perfect or exact things they would need to fix the problem. This illustrates real-world problem solving, and we do not teach this in our schools today—we certainly do not test for it. Needles to say, the scientific team came up with a creative solution, using only those items the crew would have access to onboard. They radioed their instructions to the crew, and their lives were spared.

The *real* world does not ask for the standard, "pick one of the following" solutions to life's problems. We do not need any more linear thinkers in our country working on problems that require higher-order thinking skills. The future of this country depends on *intellectual leaders*, people who graduate from high school that can not only read, write, and multiply, but also solve problems, think critically, and creatively, and anticipate the unforeseen. Only then will the future of America be secure.

CHAPTER 5

STANDARDS AND THE CULTURE WARS

Would national standards, which would require national testing, also require a national curriculum? Is this in line with the constitution? What about highly relevant but controversial topics, that may go against what children's parents believe at home, but nevertheless are important and valid theories? The previous chapter dealt with a solution to this problem through a three-part national exam, covering material set by national, state and local standards. However we delve into the Culture Wars debate if we raise these questions, but it is important to do so, to determine how assessment and especially, high-stakes standardized testing fits into all of this. (Note: the term "Culture Wars" really means, what dominant cultural theme or historical reality should we teach in our schools? White or Black or multicultural?)

Much has been written about which curriculum should be taught to our students. There are those that argue that a purely Eurocentric curriculum should be taught in every classroom in America, some believe a more Afrocentric curriculum is more appropriate, and others who fall somewhere in the middle. For the past 20 years, American schools have dealt with multiculturalism as part of the curriculum. What set the tone for this movement? In 1972, the "No One Model American" statement issued by the American Association of Colleges of Teacher Education called for an effort to support cultural diversity and global understanding. The Ameri-

The Unfinished Quest: The Plight of Progressive Science Education in the Age of Standards
pp. 71–84

can Association of Colleges for Teacher Education (AACTE) emphasized a commitment to social justice and equality (AACTE, 2005).

In the 1980s, many influential conservative writers warned of the "divisive nature" of multicultural education, and called for renewed focus on the "Western" (Eurocentric) curriculum of the Western tradition. This ideological divide sparked the culture wars. A book by Lynn Cheney called *American Memory: A Report on the Humanities in the Nation's Public Schools* (1988), was one of many stating that the encroachment of multiculturalism was a threat upon the traditional canon (as cited in Noll, 2005, p. 120). The traditional canon includes histories derived from White explorers from Europe, many whom claimed to "discover" certain American lands, even though there were others there when they arrived. Although their stories are important, American students should be exposed to the truth, and that includes the stories of those from American-Indian and African descent. But the traditional European canon writes history solely around the stories of those Anglo-European settlers who through force and disease, overtook America and proceeded to shape it as they saw fit. The European or traditional canon is vitally important for *all* students to learn and learn well, but it shouldn't be the *only* version students learn.

Noll (2005) raises some interesting debate points in *Taking Sides*, including those made by Thomas Famularo, who fears that "the multicultural movement, rather than representing diversity, is centered on the themes of race and gender and the debunking of Western culture" (p. 120). While that *is* the whole entire point of multicultural education, many cannot see through this argument, and agree with it. Those that agree with the philosophies of multicultural education, stress that diversity is a valuable resource, and that it teaches youth to be comfortable with differences and to acknowledge the contributions of others who may not be male and White. Among the accusations hurled against those who are trying to block a more multicultural curriculum, is that the real motivation is to restore more conservative themes and discredit the gains made from civil rights movements in the 1960s. Another is that the Eurocentric curriculum is warped and restrictive (p. 154).

As the economy becomes more global, it seems that our education should try to become more balanced also, since our high school graduates must meet the challenges in the workforce to successfully work with those of other races, religions, customs, and ethnic persuasions. Schools should reflect society, they should not be little bubbles of fictional reality. Even suburban schools can incorporate multicultural education into the curriculum, since many of those children will not be living in the communities where they grew up. This is a fairly recent phenomenon occurring with regularity in modern families. As children grow and leave home, they move across the country at growing rates, following jobs or opportunities. No longer will you find all of the members of one family in the same city, as you did a generation ago.

Nieto makes several good arguments for incorporating a more multicultural curriculum into the schools, among those:

- public schools are undergoing a dramatic shift that reflects the diversity of the general population.
- Most teachers are White and middle class who grew up in White neighborhoods. Many do not see any Blacks until they go to college. Practically all of them do not know Black slang, meanings behind phrases or gestures, or cultural nuances. But many of the teaching job shortages are in diverse schools in urban centers.
- Affirming diversity is all about social justice, not political correctness. We must examine *why* and *how* schools are unjust, issues that create inequity, such as social policies, curriculum, testing, tracking, and so forth. Students of color bear the brunt of this structural inequality.
- Schools show growing inequalities in these areas:

  Unequal funding
  Tracking
  SAT scores that correlate perfectly with income rather than intelligence or ability.
  By result, schools become racist and classist.

- Ignoring diversity is preparing students for a society and world that no longer exists (Noll, 2005, pp. 120-131).

Culture War Camps

One huge component of the Excellence Reform movement in the 1980 was the battle over curriculum (dubbed "The Culture Wars). The Culture Wars were brought on by dwindling United States economic influence, mass communications, and by national demographics rapidly changing. Since 1924 European immigration had been restricted, but in 1965, America opened up the gateway for Asian immigration. It is estimated that within one generation, up to ½ of U.S. schoolchildren will be either African American, Hispanic, or Asian (M. Berube, 1994).

The combatants divided up into three groups:

Western Traditionalists

Western traditionalists won the battle in the media, by arguing that the U.S. possessed a common culture derived from European origins. E. D. Hirsch's 1987 book titled *Cultural Literacy, What Every American Needs to Know*" became an important reference for this movement. It argued for a common culture after almost 2 decades of a national focus on ethnicity and diversity.

The Civil Rights movement brought to the forefront an awareness and emphasis on ethnicity. From the Progressive era to the 1960s, the emphasis had been given to Eurocentric viewpoints. But with the jolt from the Civil Rights movement and its activists, universities began to develop ethnic studies programs and departments. Multiculturalism became a whole new field. Hirsch reacted against this movement and proposed a "counter-reform" to move schools back to a more traditional curriculum.

Among those supporting the Eurocentric curriculum was Diane Ravitch, who argued strongly against an Afrocentric curriculum, calling it a bad idea. Ravitch called for a "common culture that was multicultural." Ravitch also argued that the whole reason for U.S. public schooling is to support a U.S. common culture (p. 113).

The Afrocentrists

The Afrocentrist declared a need for curricula that reflected African origins and led by Molefi Asanti and his 1980 book, *Afrocentricity*. For Asanti, the fight was not really about curriculum, but about reconnecting African Americans with their culture, which went back to their roots in Africa. Asanti called Malcom X and Martin Luther King Jr. "great prophets." Asanti also argued that the Western traditionalists suffer a "cultural lag" since the world is now so global (p. 113).

Multiculturalists

Multiculturalists, a middle group, emphasized both Eurocentric and multicultural. Even though traditionalists won the media war, schools, and colleges across the nation agreed with the multiculturalists, and began to fashion their curricula around this philosophy. Most school districts today offer some form of multicultural education, whether it be "Black History Month" in February, or the inclusion of the study of famous Black leaders through history.

One of the first major confrontations over the curriculum wars occurred in New York City in 1989, when Thomas Sobol (a White liberal) became commissioner of schools. Sobol developed a task force to "fix" the curriculum and dubbed it "The curriculum of inclusion," a term that is widely used in teachers colleges and universities in New York and across the country today. The curriculum of inclusion stated that Blacks, Latinos, and other minorities, had been the victims of intellectual oppression, and this would result in drawing up a new curriculum. As other countries caught up with the United States economically, Sobol realized that American schooling had to evolve and reflect societal changes (p. 113).

English as a Second Language

What about those students who are new to the United States, and whose primary language is something other than English? According to Darder (1991), "Study after study reveals that children from the dominant culture do far better on IQ tests than bicultural children" (p. 14). This is due to linguistic difficulties with English. Similar assessments could hamper progress for English as second language learners in the classroom. How is it possible for a youngster who speaks another language other than English, to keep up with an English speaking teacher and to make progress at the same rate as his or her English-speaking classmates? Should they be punished if they can't?

If the true nature of American democratic schooling is to even the playing field and to allow each citizen the opportunity of advancement in life through education, then how does this mentality fit in to this vision? In California, by 2014, nearly all of it's K-12 schools will likely be labeled as "failing" under No Child Left Behind (NCLB). Schools receiving Title I funds will face serious sanctions, or closure. In fact, NCLB includes 37 ways schools can fail, in many ways making it very nearly impossible for schools to succeed. Schools considered good in every other standard of measurement are branded as failing (Posnick-Goodwin, 2004). What does this do to the morale of the teaching and administrative workforce who are trying their best to teach children, many who can barely speak English? The answer is frustration and low morale, when teachers try their best, and progress is made, but the school is labeled as failing when the standardized test scores are not as high as middle class White school in the next neighborhood.

How are urban schools different than suburban schools? The U.S. Department of Education defines urban schools as those in which 75% or more of the households served are in the central city of a metropolitan area. There are currently 575 urban districts in the U.S. 11 million children attend urban schools, making up ¼ of the student population for the United States. 43% of these are classified as minority children, those of races and ethnicities other than White. In many cities, the minority student population approaches 100%. Of these children, almost all are poor or their families fall under the poverty line, and many fail to meet the minimum standards or pass the standardized tests. Most of the minority student population are not proficient in English, even if English is the child's native language (U.S. Department of Education, 1994). Garcia states (2004); "Schools are presently judged, valued and compared according to standardized tests ... the question that arises is how can these students' cognitive abilities be fairly measured using the same exam that is used with monolingual students?" (Garcia, 2004).

Standardized Testing and Race

For many years now, psychometricians have realized that certain youngsters cannot do well on certain standardized tests; not for lack of intelligence, but for the content on the tests. How do high-stakes tests impact minority and disadvantaged students? Much has been written on this subject, beginning with those who argue that high-stakes standardized tests are biased against lower-income youngsters. According to Pollard (2005), many of these tests include questions that would be answered correctly by students from more affluent and middle class backgrounds. Knowledge gained outside of school is measured on these tests, which is a major point if we want to learn the truth as to why poor children do not do well on them. For example; there are sections on some tests that deal with analogies or "what goes together" questions.

What is the opposite of:

up

(a) down
(b) sideways
(c) over
(d) under

This example is simple and straightforward, with down being the correct response. However, there are many questions like the one below, which obviously favors middle class kids:

What goes together?:

Cup and:

(a) table
(b) saucer
(c) plate
(d) floor

The answer for a middle class child who has seen his parent's cups and saucers in the china cabinet, or who has sat at a Thanksgiving table set with china, crystal, and linen, and silver, would be "b" saucer. However, to the child living in poverty, there may be cups, but certainly no saucers, so how could they possibly answer "b" saucer? The appropriate answer for a poor child might be "a" table, since that is the only thing that child has ever seen a cup sit upon. Outside life experiences are tested on these high-stakes tests, and there is no wonder indeed why poor children have more difficulty than middle-class affluent children. These tests expect those children with few advantages and impoverished home conditions to do equally well as those with every advantage. One of the main arguments many people make against giving accommodation to poor children is that poor children get the same education, since they sit right next to the White kids or go to the same school. Even in these circumstances they do not receive the same "education," since the vast majority of any person's education is received at home (Pollard, 2005).

There is much uproar concerning high-stakes tests and minorities. Harvard University cosponsored a forum called "The Civil Rights Project at Harvard and the American Youth Policy Forum," which met with a panel of prominent researchers and policymakers on Capital Hill to hear policy and implications of research on the impact of high-stakes testing on minority and disadvantages youths (American Youth Policy Forum, 2000). Several finding were discussed:

- According to Kornhaber, substantive learning goes down and rote memorization increases with high-stakes tests.
- According to Natriello, testing may be used as a barrier to graduation or educational advancement.
- According to Resnick, there is a difference between a standards-based system and a test-driven system. Standards-based systems demand expert instruction from teachers, where exams and curriculum are aligned with the standards. Test-driven systems reduce the quality of education, especially in Latino schools. Pressure on principals pushes them to sacrifice diverse education for one easily graded. Good standards-based systems are being carried out in New York and Pittsburgh.
- According to Orfield, high-stakes hurt low-income and ethnic minority students and is linked to higher dropout rates (American Youth Policy Forum, 2000). In truth, is this the answer for academic achievement? Do we truly want a whole segment of society uneducated because the standardized tests put them at a disadvan-

> tage? Is this what raising standards should be all about? Raising standards for whom?

Moore (2003) has compiled a series of both pro and con debate statements concerning the topic of whether or not standardized testing helps or hurts minority low income children in an article titled "Does the state standardized testing program help or hurt low income students and students of color?" Experts weigh in on both sides, and both have valid arguments. Kerry Mazzoni, California Secretary of Education, praises the tests, saying that they raise standards and expectations for those students, the system focuses on them, and spends money on them to ensure they succeed. Indeed, this is the highest calling of raising standards, instead of punishing students for failing, helping them to catch up. No one argues with Mazzoni on this. What many fear though, is that many systems, instead of remediation and help, those students who fail fall through the cracks, either dropping out, or giving up.

Goldie Buchanan, of the African American Parent/Community Coalition for Educational Equality, makes an interesting point when she says that Black and White children are instructed differently, and that Black children are not expected to achieve as Whites do. Suburban teachers with high credentials can help all children learn, but low income neighborhoods are five times more likely to have noncredentialed teachers, and as a result, these students have lower test scores. The tests can be used as eliminators to eliminate a child from progressing to the next step (Moore, 2003).

Kozol (2005) offers much evidence to support his claim that standardized testing unfairly hurts poor African American students. He cites the Head Start program, created 40 years ago by Congress and touted as the program that would ensure that all preschoolers would arrive at the same level and ready for kindergarten. The truth is that 40% of 3- and 4-year-olds qualifying for Head Start were denied entrance into the program. In New York City for example, in 2001, only 13,000 4-year-olds were in a Head Start Program (p. 52). The same problem exists in large cities all across America. In Boston, a first grade teacher told Kozol that a child in her class had received no preschool and was lagging far behind the others. Two years later, these children compete against those who attended the "Baby Ivies" programs (private competitive preschool). By third grade, those fortunate children have had almost 7 years of formal education, nearly twice as much as their less fortunate classmates. Yet both groups are required to take the same third grade high-stakes standardized tests (p. 53). How is this fair? How can this scenario *not* result in student frustration, "acting out," lack of ownership in school, and high dropout rates?

One of the most publicized reasons for the creation of the standardized test movement was to hold teachers accountable for their performances, blaming them in large part for the failures of children in our schools. Yet, all one has to do is to visit a classroom in any inner city school and you will find teachers doing their absolute best to teach, instill values, perform miracles every day with very little pay and materials, and dealing with students frustrated beyond measure by a system that ignores them. Instead of changing the system, which would require facing hard truths and focusing money in the right direction, it is easy to blame those in the front lines. Kozol (2005) sees this often during his research; teachers at war, battling every day for the souls of inner-city youth, armed only with the social justice and philosophy of personal contribution they learned in their progressive teacher's colleges. American teachers are blamed the most and given the least, we would not even expect firefighters to fight flames without water.

With the current "penalties for failure" mentality, one has to wonder how this affects those children who lag behind in either grade level, reading or math, or some other area, already struggling to catch up to their peers. One teacher in the South Bronx told recently of the high anxiety levels that students pick up on from teachers, struggling to get their students up to snuff for the tests. In this particular fifth grade class, the humiliation threat was high, brought on by the knowledge that their reading scores would be announced to everyone. "There must be penalties for failure" is the tone of those advocating this system (Kozol, 2005, p. 74). Anyone who ever took a psychology 101 class should know that learning cannot occur under mental duress. (In this case, abuse).

One of the most famous studies on race and standardized testing was done by Claude Steele of Stanford in 1999. Titled "Thin Ice: Stereotype Threat and Black College Students" it brought to the forefront the psychology behind the reason why intelligent middle-class Black college students failed to perform as well as Whites on a standardized test. According to Steele, the explanation has less to do with ability than the psychological threat of stereotypes about their capacity to succeed.

Steele (1999) and his associates in this study, Joshua Aronson and Steven Spencer, noticed that when middle-class Blacks attend college, they are faced with confronting stereotypes as normal or factual. Whether it is the class discussion of *The Bell Curve,* which posits as a probability the fact that Blacks are genetically inferior to Whites, hence their lower scores, or overhearing White students argue at lunch that affirmative action has let in too many unqualified Blacks, Black students hear this kind of thing all day long, every day. Steele calls this "stereotype threat," when a Black student internalized these statements as facts, and live their lives accordingly.

How did they measure this stereotype threat? Did they find that it affected academic performance? Steele and Aronson asked Black and White Stanford students to their lab and gave them (one at a time) a 30 minute verbal test made up of GRE (Graduate Record Examination) lit questions. It was a difficult test for all concerned, which was important, since Steele (1999) wanted to introduce stress. Would the Black students become intimidated and do poorly?

Even though the two groups had been statistically matched by ability level, the Blacks did significantly poorer on the test, so something other than ability was to blame. So they returned to the lab. This time, they told the students that the test was going to be used as a study to see how certain problems are solved, *not* a test measuring ability (which Steele stressed to the students). This changed the outcomes dramatically. The stress levels were "turned-off" and the Blacks performed exactly the same as the Whites (Steele, 1999). There cannot be a more dramatic demonstration of the psychological contribution to the harm standardized tests can do to minority students who have grown up hearing how inferior they are to Whites, and thereby internalizing these misconceptions.

Tracking

Why are there so few African Americans in advanced placement physics classes? There is one theory: tracking. Acceptance of diversity may also reign in the insidious practice of tracking: the placing of certain people into groups or educational tracks, to where the present placement in most cases determines which educational choices and future will be available to the "trackee," in essence, making very early determinations of no less than the student's fate in life. The "track" more times than not, is awarded according to racial or socioeconomic status, rather than by ability. I once knew of a little Caucasian girl, tracked into the "slow" math class in the 1960s, only because her parents were going through a divorce and the family was temporarily on public assistance. (At the time divorce was rare in the suburbs). Instead of delving into the girl's out-of-character withdrawal and quietness, the teacher placed her in the lower-achieving group based on outward appearances, which was to follow her throughout her schooling and took great effort to overcome. The girl turned out to be a superior student who went on to get her PhD.

Tracking students by "ability" is nothing new. It is been around for generations. What is bothersome about tracking, is how globally accepted it is as a valid form of education. If one views tracking on the surface, one can assume that ability grouping is nothing more than placing students with similar abilities into groups, so as to better their chances of academic success by neither forcing them into learning situations that would frustrate them, or bore them. Sounds good.

The insidiousness of tracking is what results from it. According to Tyler (1999), separating children according to academic ability really separates them according to race, forcing a new kind of segregation. Welner (2003) suggests that tracking is supposed to provide the "trackees" with differentiated instruction, the ultimate goal being to raise those on the lower track so they can catch up with their peers. The reality is that tracking is a form of "second-generation segregation"; in effect an effort to resegregate Blacks from Whites after courts forced desegregation at the school level. Indeed, this practice continues in many school districts across the country. This especially happens when schools pull populations of students from either side of the tracks. Wealthy parents who have power in these schools, can threaten to pull their children out of the school if tracking is eliminated. It truly is a way to keep children separate in the same school.

Administrators can meet with resistance if they try to de-track their schools. Carol Burris, principal of a high school in the Rockville New York school district, successfully eliminated tracking from her school. She met with resistance from White parents at first, but assured them that if it was not successful for everyone in the school, the school would reconsider. Participants in the de-tracking noticed many positive changes in the school, including increased test scores and achievement, better self-esteem among minority students, teachers improving their methodologies, decreased levels of discipline problems and violence, and better race relations (Burris, 2003).

The Civil Rights Project at Harvard University, led by Dr. Gary Orfield (2005), commissioned a study called "Testing, the Needs and Dangers" which addressed civil rights concerns associated with high-stakes tests. Among the finding of the study were these points:

- If test results are related to important decisions and outcomes, then teachers often begin to "teach to the test." A National Science Foundation study in 1992 showed that teachers with a high percentage of minority students were significantly more likely to state that standardized tests affected their teaching style. Thus, there arise serious differences in curricula between classrooms with high and low percentages of minority students.
- High-stakes tests do not necessarily make teachers and students more motivated in the classroom. Psychological studies have shown that motivation is highly complex and that people deal with it differently. Those students who are not motivated by the tests will

begin to feel alienated by the tests and consequently, the educational process.

- High-stakes testing is correlated with high drop out rates. Researchers from the State University of New York found that 9 of the 10 states with the highest drop out rates use high-stakes testing, while none of the 10 states with the lowest drop out rates do. Minority and low-income students are more likely than others to attend schools that use high-stakes tests.
- Dropout rates have risen in the past few years, especially for African American males. The costs for dropping out are high; those that do not graduate have no chance for college and little chance of finding a decent job. There is also an increased likelihood that they will be imprisoned.
- A student's performance on a high-stakes exam is significantly tied to the level of their teacher's experience. Minority and low-income students tend to have teachers with the lowest amounts of experience and are therefore likely to perform less well on high-stakes tests than their White counterparts—and to be unfairly hurt by the test's consequences.
- There is little evidence of a correlation between high test scores and job success. The test score gap between Black and White males has narrowed by half since the mid-60s, while the Black-White wage gap for males that narrowed primarily during the period of civil rights enforcement has grown since that time.
- The use of high-stakes testing results to make employment decisions is likely to be ineffective as well as harmful to minorities.
- Test-based grade retention does little to improve learning. In addition, it is expensive and may present class management problems associated with having an older student in a classroom with younger students.
- Grade retention disproportionately affects African American males who are the most likely ethnicity/gender group to be held back. (By ages 15 to 17, close to 50% of African American males compared to about 30% of White females are below the average grade for their age or have dropped out of school.) (Orfield, 2005).

Orfield (2005) also states that urban schools have slipped back into a desegregated state not seen for 30 years. Racial isolation has increased, largely unnoticed by the general public who do not keep track of public school demographics (except where it concerns them in their suburban school divisions).

> Almost three fourths of black and Latino students attend schools that are predominantly minority ... more than two million, including more than a quarter of black students in the Northeast and Midwest, "attend schools which we call apartheid schools" in which 99 to 100 percent of students are nonwhite. (Kozol, 2005, p. 19)

Before urban schools began this backward slip into segregation and were more desegregated, minority high-school graduation rates increased sharply and racial gaps in test scores closed significantly until the 1990s when segregation began again. This was due in part to the standards-based reform and school choice—where White parents could take their children out of mixed race schools and place them in more homogeneous environments. There are benefits of segregation, including better teachers, more supplies and materials, better buildings. The more segregated the school, the higher the concentration of poverty and racial isolation (Kozol, 2005, p. 20).

In September 2005, Norfolk Virginia, a large urban school district, won the nation's top award for urban school districts; the Broad Prize, awarded by the Broad Foundation, a Los Angeles-based nonprofit organization. It had been runner-up the 2 previous years. Norfolk's school age population is 37,000, the smallest system in the running, beating out school systems like Boston, San Francisco, Houston, and New York, for a total of 82 school districts. The award addresses and recognizes high achievement and improvements in the academic performance of school districts with a high percentage of minority and low income children. (African Americans make up 63% of Norfolk Public Schools) (Jeter, 2005; Norfolk Public Schools, personal communication, September 21, 2005).

If one analyzes why Norfolk won, one needs to know the history of Norfolk's educational system. In 1986, Norfolk city leaders and the school board dismantled cross-town busing for racial purposes, ending a great deal of "white flight" from Norfolk into suburban Virginia Beach, among other places. This plan was challenged by various Civil Rights groups, but the decision to end busing was upheld. As a result, slowly over several years, Whites began to move back into Norfolk neighborhoods, causing in turn a "rebirth" of the downtown area and a new vitality of the city in general. However, many schools became segregated once again. The positive result was that since there were now more middle-class Whites and Blacks moving back to Norfolk, the schools began to reap the rewards of middle-class teachers and families living near the schools. More White parents placed their children in "neighborhood" schools, in many cases bringing the racial make-up of schools to 50% Black, 50% White. The downside is that many schools remain segregated, needing funding from the Federal government to make ends meet (M. Berube, 1994, p. 60).

The ending of "white flight" from the suburbs back into Norfolk most certainly helped bring up test scores and achievement, since the middle classes of both races came back into the city. While cities like New York have sharper distinctions between richer and poorer neighborhoods, and where there it is less likely for races to live in the same neighborhoods due to costs of living, the separations between Black and White school achievement remain more distinct.

As long as those in power have a lack of understanding of what is happening in our inner city schools, they will continue to blame the victim for their plight. No one advocates a free ride, or handing someone a diploma that doesn't deserve it. But when you have the facts brought to light by many years of research by the leading social scientists of the day, the problems and solutions become crystal clear. Equal educational opportunity is all anyone wants or needs. *Access* to free public preschool. *Access* to textbooks that are the latest editions and that are not 15 years old. *Access* to teachers supported by proper pay for important work.

The issue here is not that we lower standards, or that we have no standards, or that minorities are left out of the standards process. The issue is that we take the whole learning experience into consideration when we test children on their knowledge, knowing that the home environment counts for more than half of a child's learning experience. When we use tests as punishments for being born into the wrong household instead of as learning tools, everyone loses.

CHAPTER 6

ASSESSING INDIVIDUAL DIFFERENCES, DISABILITIES, AND OTHERWISE

Standardized testing philosophies assume that everyone learns the same way, and has the same abilities. This is simply not true. What happens to bright children who do not fit into the traditional category of those who do well in either language arts or math? Do these children get a chance to succeed in school, to go to college? Are we neglecting a whole section of our youth as we shuffle them through the system?

What about children with disabilities? How do they fare with standardized tests? Morrison (2003) argues that policymakers advocating standardized testing must guard against unfairly denying educational opportunities to students in an effort to set higher standards for the general population. Some schools actually "excuse" some students with disabilities from the standardized tests in order to raise scores for the school, even though the Individuals With Disabilities Education Act of 1997 (IDEA, 2005), mandates that students with disabilities be included in the assessments and are supposed to provide the appropriate accommodations when necessary (Morrison, 2003). The National Research Council (NRC) estimates that about 10% of American schoolchildren are counted in this population; this is a very high number of students that are not being assessed according to law (NRC, 2005).

The Unfinished Quest: The Plight of Progressive Science Education in the Age of Standards
pp. 85–97

Specifically, an NRC report published in 1999 identified these conclusions concerning students with disabilities and high-stakes tests:

1. Disabilities can lead to unpredictable distortions in test scores.
2. Some accommodations may inflate artificially and inappropriately the scores of some students.
3. The most common accommodation, providing additional time, is not appropriate in every case; moreover, the effectiveness of this accommodation merits more research (NRC, 2005).

My argument has been that we not use tests as punitive assessments that may hold a child back *unfairly*. If a child does not study, do his work, or go to school, then he will obviously fail the test. But we cannot assume that children who fail do so because they are to blame.

My brother, a middle school history teacher in Virginia Beach, Virginia and a highly gifted individual, failed first grade because he had an undiagnosed case of dyslexia. This learning impairment affects how the brain decodes written language, among other things. A *CNN* news report, "Exit Exams Deny Diplomas" from Sunday, June 15, 2003, tells the story of Karl Kearns, a senior at Burke high school in Boston, who was one of 4,800 seniors statewide who did not pass the standardized test that year. It was the first year that students statewide had to pass the Massachusetts Comprehensive Assessment System (MCAS) exam. or they could not graduate. While Kearns held a "B" average (above average work and effort), and won an award for most improved in his class, his combination of a reading disability and troubles at home kept him from graduating with his friends. He had taken the exam four times and was only two points shy (CNN.com education, 2005).

On February 23, 2005, after a 10-month study, a bipartisan panel reviewed the No Child Left Behind Act (NCLB) as it pertains to IDEA. The panel asked Congress to recognize special challenges disabled students face with national testing. The report concluded that NCLB conflicts with provisions of IDEA, which required IEPs (individualized education plans) for all children with disabilities. Under NCLB, a disabled eighth grader must take an eighth grade exam, even if she is only capable of sixth grade work. This requirement for grade-level testing violates the mandate of IDEA that students be held accountable only for level of ability (National Conference for State Legislators, 2005).

O'Neill, Wilkey, Farr, and Gallagher (2000) goes so far as to question the legality of standardized testing for children with disabilities. He poses five things to look for when high-stakes tests are given to learning disabled kids:

1. Are learning disabled children being excluded? All children must be tested, but maybe accommodations or changes can be made to the test. (Learning disabled refers to children who suffer from a range of conditions that affect a person's ability to learn new information.) (Wikipedia.com)
2. Are there alternatives to the testing? What if a student cannot write but could give an oral exam? Some states offer more than one diploma, some where no passing scores on the tests are needed.
3. Has there been enough "lead-time"? This is time leading up to the actual test that counts, several years if necessary.
4. Do the schools actually teach what they are testing?
5. Are appropriate accommodations being given?

The Americans With Disabilities Act (ADA), signed into law July 26, 1990 by George H. W. Bush, gave Americans with disabilities rights and protections to ensure full participation in community services, including schools. Specifically, ADA has implications for student participation in educational programs and assessments, including the creation and administration of assessments for such students. Regulations prohibit the use of "criteria or methods of administration which have the effect of subjecting individuals to discrimination on the basis of a disability and services" (Morrison, 2003). This language ensures that special needs students not be left out of the testing process, but also implicates the need to alter or revise such assessments, and each state has the final responsibility for setting educational policy for assessments, while being held accountable by Congress.

One of the keys towards solving this problem belongs to those placed in charge of creating the assessments. How do we currently know that special assessments aimed at disabled or special needs children properly assess achievement in terms of validity and reliability? There is not enough research to address how such modifications may affect the outcomes on such tests. Test developers should include special needs and disabled children in the field tests and pilots of such assessments, to ensure validity and reliability. Only then will we be sure that these children are not falling through the cracks ("Appropriate use of," 2001).

Gardner's Multiple Intelligence Theory and Alternative Assessments

The current standardized testing movement would make more sense if all of our students were identical in terms of learning styles and abilities.

One of the leading authorities on educational differences in children is Howard Gardner of Harvard. In justifying alternative assessment, Gardner (1983) posits that there are at least seven types of intelligence to be found in schoolchildren. He names these as:

1. linguistic (speeches, writing, poetry, verbal skills)
2. musical (singing, composing, playing instruments)
3. logical/mathematical (problem solving, computation, logic)
4. spatial (artist, sculptor, architect, map/directional skills)
5. bodily/kinesthetic (dancer, athlete, ballet)
6. interpersonal (people skills, getting along with others, empathy)
7. intrapersonal (self-knowledge, emotional intelligence)

By today's standards, only two of these areas are measured by standardized tests, linguistic and logical/mathematical. Not surprisingly, these are the two areas measured on all standardized tests, including SATs which either open or close the gate to college. Traditionally, the other abilities have been termed "talents" rather than "ingelligences." One cannot help but wonder why some are called talents, (which carries with it less prestige) rather than intelligences. It is important to note here that Picasso thought himself stupid and a failure in school as he did not perform well in a traditional school setting, even though he possessed superior spatial abilities that resulted in some of the most brilliant artwork known to humankind.

Some people possess different ways of thinking, not fitted to the norm as we know it. Some of these differences can be called "disabilitites," while others can be called "genius." Some can argue that there is little difference between the two, while in other cases, there is a vast noticeable gap separating them. The "accountability" movement is not concerned with these students, or how properly to asses them. In either case, both groups are included in the cattle-herding mentality of public schooling, however, the law requires school systems to differentiate and assess all students, and to fit an educational plan to each child's different strengths. Standardized testing fails to do this. Alternative assessments measure students' understanding of content more thoroughly and completely than multiple-choice retention tests (Armstrong, 1994; Gardner, 1983).

Personal Case Study and What Autism has to do With Science

The changes called for in instructional practices require an adjustment in the types of assessment tools used to evaluate learning. It would not be a coherent strategy to ask students to perform a wide range of high level learning experiences and then measure their progress solely on the basis of standardized multiple choice tests. In the same vein, it would not be fair or wise to ask students with disabilities to perform the same as those students without disabilities. One problem with results-driven assessments is that the process is overlooked in favor of the product. What about those children who are outliers on either side of the bell curve—those who are either extremely bright, those that fall below the average, or those who just might possess exceptional qualities on either side? The problem is this: what a school system views as a disability in this product/grade-driven world, may turn out to be valuable learning style difference.

The disability that comes to mind is autism. In my opinion, this disability is a perfect example of how traditional schooling can overlook individual gifts. Those colleges of education that face the future well prepared are those that begin studying autism and how best to educate this population of children who seem to be literally exploding onto the scene at an astounding rate. The future of American education will depend on special education teachers and how well-prepared they are to handle individuals with Autism, and individual differences, while encouraging individual strengths. I would like to make the connection here between autism-spectrum disorders and Gardner's (1983) intelligence called spatial ability.

It is everywhere lately—books, television, magazines, and newspapers, Autism and related disorders command the headlines. According to the NIH (National Institutes of Health), there has been an increase in true diagnosis of autism of about 30-60 per 10,000 population, up dramatically from the figure 40 years ago of four per 10,000 (Rutter, 2005). Everything from immunizations, allergies, genetics, and intestinal problems have been blamed for the rise. Complications occur also when doctors diagnose children with autism when they have another related disorder. This can happen when a diagnosis is attached to funding involved for treatment, or to placate parents who won't leave without a definite answer from the doctor. For whatever reason, the numbers have skyrocketed. But if true autism accounts for a small number of recently diagnosed cases, what exactly do the others entail, and must it always have a negative connotation? In other words, do the characteristics of many autism related disorders have to be a bad thing?

At this point I have to clarify some definitions. True traditional autism involves severe learning disabilities that may include all, some or a combination of speech delay or absence altogether, repetitive or stereotypical behaviors (spinning objects, rocking), disconnection from humans, low eye-contact, sensory sensitivity or sensory craving, discomfort with touch, sometimes mental retardation, and many other various afflictions. (Before the media blitz, most of us knew autism only by the Dustin Hoffman character in *Rain Man*, an "idiot savant," who exhibited most of the signs of true autism, but who possessed a gift for memory and math. This is extremely rare.) True autism is just one category that falls under the "umbrella" of pervasive developmental disorders (PDD). These disorders fall on the "spectrum" of disability, and can include speech delay, motor delay, one or several of the symptoms, and in varying forms and degrees. Some children diagnosed with PDD are hardly afflicted, and it would be hard to tell how they are different from other children, others are severely afflicted and fall under the diagnosis of true autism. In any case, the numbers for all related disorders are growing at staggering rates.

I began my journey into autism recently, but I have always been fascinated with how seemingly unrelated concepts can be connected. Six years ago, I wrote a piece titled "Multiple Intelligences and the Artistic Imagination: A Case Study of Einstein and Picasso" (Newbold, 1999). In this article, I used Gardner's (1983) multiple intelligence theory to argue that Einstein and Picasso had similar thought processes, even though one was a scientist and mathematician, (so-called "left brain" person) and the other was an artist, (or "right-brain" person). Since then, I have given birth to a little boy who is 6 years old at this writing. He happens to be a late-talker. When he approached his second birthday, his father and I became concerned when he did not seem to be acquiring the language appropriate for his age, so we started him in speech therapy. After our move to the New York area, he was still cheerful and curious, and could understand what we told him, but was still lagging far behind verbally and socially, so we had him evaluated by a pediatric neurologist and another team of experts. He was given many diagnoses; the neurologist told us that he did not fit into the autism category, although he exhibited a few symptoms, but not enough in each category. The other team told us that they were hesitant to apply this label, but that he could fit under the "autism spectrum" umbrella of PDD-NOS, which stands for pervasive development disorder, not otherwise specified. Their term for him was "atypically atypical." His most recent diagnosis, from an autism specialist, was that he did not have traditional autism, but a communication disorder on the "autism spectrum," which can mimic autistic symptoms until the communication becomes better. He definitely does have autistic symptoms, though not others. Needless to say, his father and I were confused and decided to do some research ourselves.

I thus began researching PDD-NOS and autism related material. The more I read, some material written by doctors, some by autistic grown-ups, I began to see Autism in a new way, not surprising with my penchant for making connections between seemingly unrelated areas. Keeping in mind that most disorders are "spectral," meaning that there are levels of disability each person with a disorder can have. Some are high-functioning, some are severely debilitated. It was then that I stumbled onto the writings of Dr. Temple Grandin, who is the foremost authority in the country on humane animal science. She is a livestock handling designer and associate professor of animal science at Colorado State University, who developed a revolutionary system of handling and transport of cattle either on their way to see a veterinarian, or to slaughter, which focuses on humane handling and reduced pain and fear for the cattle. No on in the country ever thought of this before Dr. Grandin. Facilities she has designed are located across the world, and almost half the cattle in North America are handled in a center track restrainer system she designed for meat plants (Grandin, n.d.). She is also autistic.

Temple Grandin's story begins as a little girl, where she says that she didn't talk until she was three and a half years old, and communicated by screaming and humming. Her parents were instructed to put her away into an asylum, but she was instead sent to a series of private schools, finally majoring in experimental psychology and obtaining masters and doctorate degrees in animal science (Encyclopedia Britannica Online, 1999). As a teenager, she became aware of feelings of increased anxiety, and devised a system to alleviate it. It became known as "the squeeze machine". She modeled it on a chute fashioned to hold animals in place during branding and other procedures (Encyclopedia Britannica Online, 1999). The squeeze machine applied pressure to her body, and the feeling of pressure and tightness comforted her anxiety symptoms. The alleviation of her anxiety compelled her to try to design a system that would reduce fear in cattle as they were led to slaughter. She credits this talent to her astonishing ability to see in pictures.

Dr. Grandin (n.d.) describes the ability herself:

> One of the most profound mysteries of autism has been the remarkable ability of most autistic people to excel at visual spatial skills while performing so poorly at verbal skills. When I was a child and a teenager, I thought everybody thought in pictures. I had no idea that my thought processes were different. In fact, I did not realize the full extent of the differences until very recently. At meetings and at work I started asking other people detailed questions about how they accessed information from their memories. From their answers I learned that my visualization skills far exceeded those of most other people. (Grandin, n.d.)

This was an excerpt from her book, *Thinking in Pictures* which she wrote in 1995, in which she describes how she can learn nothing that is language based, a traditional standardized test skill. She converts things in her mind to images in order to understand them, a very unusual way of thinking. She performed this mental exercise throughout school and even through college. This visual thinking has helped her enormously in her development of designs for the livestock industry. Visual thinking is also one of Howard Gardner's "multiple intelligences" (Gardner, 1983). This form of intelligence, which Gardner calls "spatial", enables a person to base his or her method of learning on visual thinking instead of language-based thinking. According to Grandin,

> I think in pictures. Words are like a second language to me. I translate both spoken and written words into full-color movies, complete with sound, which run like a VCR tape in my head. When somebody speaks to me, his words are instantly translated into pictures. Language-based thinkers often find this phenomenon difficult to understand, but in my job as an equipment designer for the livestock industry, visual thinking is a tremendous advantage. (Grandin, n.d.)

Indeed, many parents of autistic children know very well that the only way they became potty trained was through picture related promptings. Visual learning is how they make sense of the world.

It is fascinating to read of Grandin's accounts of how she discovered what scared cows on their way to slaughter—the edge of shadow, certain bright lights, that no one up till then had thought or cared about. She discovered this and other amazing things by bending down and traveling through the chutes to see the world through a cow's eyes. It was during college that she began to realize that she thought differently than other people. She remembers reading an article by a famous scientist that stated that early humans had to have invented language before they could develop tools. She was amazed at the ignorance of this statement; she had never used language in the development of all of her inventions, yet here was a scholar in the field stating it as if it were fact.

Many people who study autistic children think that they lack an imagination because the adults do not see any evidence of "meaningful play," indeed, supposedly one of the "symptoms" of autism is that children show no interest in imaginary play. This is inaccurate. According to Grandin, her whole method of assembly and "dry-runs" of her inventions are carried out via her imagination, other people around her unaware of the brilliance taking place right next to them.

> Today, everyone is excited about the new virtual reality computer systems in which the user wears special goggles and is fully immersed in video game action. To me, these systems are like crude cartoons. My imagination works like the computer graphics programs that created the lifelike dinosaurs in Jurassic Park. When I do an equipment simulation in my imagination or work on an engineering problem, it is like seeing it on a videotape in my mind. I

> can view it from any angle, placing myself above or below the equipment and rotating it at the same time. I don't need a fancy graphics program that can produce three-dimensional design simulations. I can do it better and faster in my head. (Grandin, n.d.)

Arguments are made that Einstein too, was on the "autistic spectrum," albeit a high-functioning member of that group. (He did not speak until he was 5 years old. His father was told by teachers that he would not amount to much and was probably retarded.) As mentioned earlier in this book, Einstein developed his theories for relativity as a child by visualizing himself riding a beam of light. His visualizations were amazing because he could "see" whole complete theories in his head, much as Dr. Grandin does. One of the startling discoveries one can have of Einstein is that he failed many math courses and found them dry and boring. He was also a poor speller and equally poor in foreign language. This visual ability was my defense for my proposition that Einstein thought like an artist instead of a mathematician.

Thomas Sowell (1997) has written two of my favorite books on late talking children, one titled *Late-Talking Children* appropriately enough, and the sequel, *The Einstein Syndrome: Bright Children who Talk Late*. In both of these books, Dr. Sowell speaks from experience of his now college educated son who at age three couldn't talk and who other people labeled as retarded. Sowell knew his son was bright though, and finally, right before he turned 4, started to speak in sentences. Now his son, John Sowell, has a degree in statistics and creates his own computer games. The book is titled *The Einstein Syndrome* because of the reference to Einstein's late speech development. Indeed, three of the people responsible for the atomic bomb, were very late talkers, including Einstein, Edwin Teller, and Richard Feynman (Sowell, 2001, pp. 40-47). While I hope my son falls in this category of outcomes in late talking children, no one really knows what the future will hold. Many autistic children do have limits in cognitive ability, while many do not, and are in fact gifted in one way or another. Those labeled PDD-NOS are less likely to have mental deficits than those labeled as classically autistic. The question arises, at what level can *high* intelligence be though of as abnormal?

In one of Grandin's most interesting articles (http://www.autism.com/families/problems/genuis.htm), "Genius may be an Abnormality," she writes that genius in any field may be an abnormal state. There are two-times as many engineers and computer programmers in the family history of people with autism than those without, and Grandin is no exception. Her grandfather was an engineer and coinvented the automatic pilot for the airplane. She contends that a little bit of autism genes provide an intellectual advantage, while too much may cause severe impairment (Grandin, 2003).

So, as time goes by, in terms of Einstein, theories of spatial ability fueling the greatest mind known to science, are confirmed and strengthened. Can this "top-down" inductive way of thinking really be the secret behind the greatest inventions and discoveries of our time? Can the ability to see the whole, complete vision first while plugging in details later, equip someone with the talent to solve problems most of us can only wring our hands over? Another subject of my intellectual admiration is a professor of physics and mathematics at Columbia University, Brian Green (2000). Dr. Green has given the general populace the gifts of his two books, *The Elegant Universe*, and *The Fabric of the Cosmos: Space, Time and the Texture of Reality*. In both of these books, Dr. Green uses visual prompts and drawings to help us understand the newest attempt to incorporate quantum mechanics into Einstein's theory of relativity with string theory (or superstring theory). While not the inventor of this theory, he has become it's poster boy, with mesmerizing specials on PBS, books and videos describing in amazing visual graphics, the tenets of string theory. Green uses visual images, with little math to frighten off the person learning these theories for the first time.

Now, Brian Green is not in any stretch autistic, but he also possesses the gift for visualization, which is extremely helpful for those of us who need to "see" something in order to understand it, if only in our minds. Dr. Green is a rarity in the science field; a mathematician and scientist who is highly verbal and analytical at the same time, hence the popularity of his books.

What does Brian Green have to do with children who may be autistic? I believe that there is a link between visual thinking/spatial ability and intelligence, and autistic or not, people like Dr. Green and Temple Grandin have this trait in common. In Dr. Green's case, he also possesses superior mathematical and verbal abilities, and is able to perform very well on standardized tests. But there are many people, like Picasso, who possess the brilliant spatial abilities, but cannot perform well on these tests. Most people on the autism spectrum are visual thinkers. In many cases, augmented communication can help autistic individuals when nothing else could. Toilet training picture cards have helped many autistic children potty train because they understand the pictures before they understand the words involved with potty training. It is a different way of thinking and processing information, but it adds a unique viewpoint and one that contributes to our understanding of the world, rather than detracts from it. Visual intelligence is an enormously important talent, and it needs to be cultivated in those who possess it, autistic or not (C. Berube, 2007).

Areas of Brain Function and Implications for Autism

It is helpful to know what parts of the brain are responsible for speech and communication when studying Autism related disabilities (Figure 6.1). Many questions come to mind when considering the mystery of speech delay in autistic children. One that sticks out for me is the obvious locations of speech areas in the brain in the left hemisphere, and the area for spatial abilities in the right hemisphere. Most autistic adults who can now communicate state that they could understand language as children far earlier than they could respond verbally or communicate, thereby causing the acting out and tantrums due to frustration. Nonverbal communication, such as gesturing, grunting, and other forms short of speech were also employed before full language was acquired. It is curious

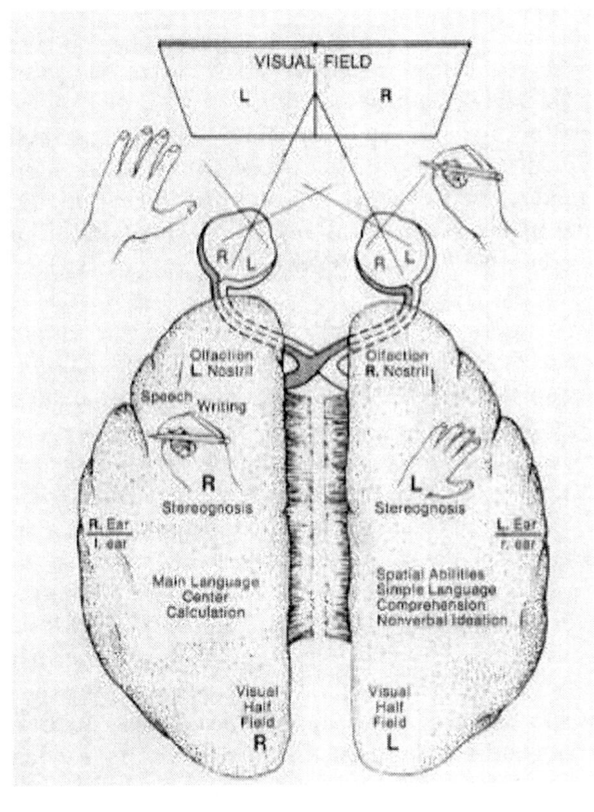

Source: Diamond and Scheibel (1985).

Figure 6.1. Areas of brain function and implications for autism.

to see how these specific areas of the brain are controlled by such functions, and how the two hemispheres of the brain must communicate with each other in order for language development to occur. One theory in the mystery of Autism might be that there is a short-circuit between the left and right hemispheres of the brain, that would enable normal speech to develop. It is also important to remember that most autistic children and adults possess superior spatial skills than does the normal person. Also interesting, is the fact that most autistic children have trouble with writing skills and gross motor skills in general. Consider this definition from Encyclopedia.laborlawtalk.com of dysgraphia:

> **Dysgraphia** is the inability to write, regardless of ability to read. People with Dysgraphia often can write, but lack co-ordination, and find other tasks like tying shoes difficult. They also lack basic spelling skills, and often will say the wrong word when trying to formulate thoughts.

If one considers the area of the brain responsible for writing skills, it is noticeably near the language center, which makes one wonder if some dysfunction globally on the left side of the brain is the culprit? If Temple Grandin's article "Genius may be an Abnormality" is accurate, does this explain why many extremely bright people possess poor handwriting? Is genius the mildest form of autism disorder or related in any way? It will be fascinating as the years reveal answers to these and other questions parents and educators have about developmental delays and brain development.

With this in mind, hopefully the future of autism research will help develop ways to properly study and facilitate brain function, while maintaining the individual advantages of superior spatial ability in such individuals, as was the case with Temple Grandin.

When I wrote *Multiple Intelligences and the Artistic Imagination: A Case Study of Einstein and Picasso* back in 1999, I had no idea of the avalanche of scientific evidence that was soon to come to support my theory of spatial intelligence and how two such seemingly polar opposite geniuses possessed the same gift. I also had no idea that I would give birth to a special child that would change the way I view the world and make me delve deeper into intellectual differences and learning style research.

In our growing understanding of autism and spectrum disorders, if we look at them as something to totally cure, we will miss the point. Grandin points out that children with these disorders need intense therapy to pull themselves out of their world into the outside world of interaction. She encourages those with autism to pursue jobs that have nothing to do with autism, to exercise those parts of the brain that force one to function in the world. She also says that speech, behavioral and occupational therapy are necessary to focus them but not to "cure" them, because to do so would be to commit a kind of genocide against a culture of people that may be in possession of gifts that could literally, change the way everyone views the world, whether it's revolutionary advances in bovine science, or space and time. Maybe even our definition of reality (Berube, 2007).

Awareness of differences is not enough. As stated earlier in this chapter, we must apply this knowledge to the assessment process if we are to have a truly fair, democratic school system. This case study might seem an extreme example of an argument for different forms of assessment, but if trends continue, there will be an explosion of students with various pervasive development disorders in our schools in upcoming years. Alternative assessments are more expensive and time consuming to perform than traditional multiple-choice standardized assessments. They require training and knowledge on the part of the administrators of such tests. The question is this: do we want to invest the time and money into properly assessing *all* children, as is required by law, even those with disabilities, delays, and *abilities* that are not the "norm?" How do we assess those children not in the middle or high end of a bell curve? What about gifted children whose answers often do not match the required instructor's answers, but are on a higher level of understanding and creativity? Under the current assessment system, we are punishing creativity while rewarding the "herding mentality" of education, where all converge upon one singe reality; the correct answer.

CHAPTER 7

SEXISM AND STANDARDS

Once again we cannot escape the truth of differences. Do boys and girls learn differently? If they do, what are the implications for instructional practices, and for assessments? In terms of assessments, how do high-stakes standardized assessments affect girls? May dangers lurk in this pool of educational research, because there is a tendency to equate difference with one being superior to the other, when in reality, different can just mean different methods of achieving the same goal, or different ways of taking in and responding to information, both being valuable and valid.

Carol Gilligan (1993) conducted important research detailing the differences between boys and girls learning styles and girls' ties to self-esteem. Gilligan's work is relevant to education because it studies how girls thinking patterns affect learning. Girls think in "circular" ways, whereas boys tend to think in a more linear fashion. One is not superior or inferior to the other, and is mainly due to brain wiring, but both require different means of assessment. Scott Brewmen (2003) offers this information:

> "Size Does Matter
>
> The reasons behind these differences (between men and women) have fuelled arguments for generations and continue to do so today. Is it our biology, or our culture? Many scientists say it is all in our heads, or, more precisely, in the way men and women's brains are designed and the way they function.

The Unfinished Quest: The Plight of Progressive Science Education in the Age of Standards
pp. 99–105

> A century ago, the discovery that female brains were about 10% smaller than male brains was cited as proof that women could never be as smart as men—contributing to their status as second-class citizens. We now know that size isn't everything when it comes to the power of brain. Our IQs are the same. In fact, the highest recorded IQ belongs to a woman, a writer named Marilyn vos Savant.
>
> There are other, perhaps more significant, differences that distinguish male and female brains. Male brains are wired to move information quickly within each side—or hemisphere—of the brain. This gives them better spatial abilities. They can see an object in space, and react quickly. In women's brains, areas of the cerebral cortex—linked to language, judgment and memory— are more densely packed with nerve cells than men's brains. This allows them to process that information more effectively.
>
> Fisher explained that the "corpus callosum," which she describes as a "big highway between the two sides of the brain," is larger in women toward the rear than it is in men. "Hence," she said, "the two sides of the brain are better interconnected" in women. This means that women can absorb and analyze all sorts of information from the environment simultaneously. This makes women more adept at multitasking, while men tend to do better tackling one thing at a time. (Brewman, 2003, para. 5)

Circular versus linear. In order to illustrate this circular pattern of thinking (out of the box is a more modern term), Gilligan (1993) posed the "Heinz dilemma," a hypothetical moral problem that highlights the different ways males and females solve problems, to two 11-year-old children, one boy and one girl, in the same sixth-grade class at school. In this moral dilemma, a poor man named Heinz considers whether or not to steal a very expensive drug from a pharmacy in order to save the life of his sick wife who needs the drug to survive. The problems are, his wife is sick and the druggist refuses to lower the price of the drug. The question is, should Heinz steal the drug?

The boy answers that yes, Heinz should steal the drug, because after all, it would save the life of his wife, and that is more important than stealing in this case. A valid reason, based on logic. The girl comes up with a different answer that illustrates how girls think differently than boys. She says that Heinz should not steal the drug, but instead should try an alternative solution, either bartering with the druggist about if Heinz could maybe work in the drugstore to help pay for the drug, or borrow the money or take out a loan. Her reasoning was that if he was caught stealing the drug, he could go to jail and then his wife would really be in trouble (an interruption in the relationship). These different solutions speak to linear versus circular thinking—the boy's answer followed a very logical pattern, the girl's answer took into account relationships and how best to preserve them, including the relationship between the druggist and Heinz, and between Heinz and his wife. The girl saw the solution as less of a "math problem with humans" than as a narrative of relationships that extend over time (Gilligan, 1993).

The ways in which girls view groups or webs of relationships as being the most important aspect of their lives affects their learning styles and schooling and is reflected in data that shows that girls learn best in cooperative learning situations (Gilligan, 1993). Research abounds with accounts of girls who, although superior to boys in science in elementary school, somehow disengage by middle school to the point where almost no girls occupy spots in advanced placement science classes in high school. If girls show that they can hold their own in the science classroom, they run the risk of being "cut off" socially not only by boys, but by many of their girl friends as well. The result is that there are two children, one male and one female, both highly intelligent and perceptive, though in different ways, with different conceptions of understanding the world.

Cooperative Learning and Standards

Some subjects more naturally loan themselves to cooperative group learning. Science for example, (at least the lab portion anyway) is a social subject by its very nature, and children have to work together in lab groups and cooperative situations. Science classes are wonderful examples of Vygotsky's social learning in practice. Cooperative learning comes naturally to most girls, since girls assume human connection naturally. To be more specific, girls begin to experience separation as a *new* experience as they grow, having formed intense bonds early on, whereas boys assume separation very early on and begin to experience connection as their new experience (Bailey, 1992). This drama is acted out in every classroom in America.

This theme of being abandoned socially, is Gilligan's (1993) other main theme of her work; that girls value connecting and webs of relationships at all costs, where boys value just the opposite: disconnecting and individuality. American schools are set up to reward competition that grows out of individual accomplishments, not cooperation. Girls will submerge their intellectual ability in order to fit in to their highly prized social group in order to belong. As a result, science and math are the first scholastic subjects to show the effects. In short, Gilligan brought to light how girls bring their own meaning to situations that may be different from their male counterparts and originating from different life experiences, once again a constructivist notion.

Cooperative learning is beneficial to all students, but especially for girls, whose inherent propensity for social learning fits perfectly with group problem solving, and the verbal nature of cooperative learning.

The verbosity of cooperative learning promotes problem-solving and alternative solutions, both higher-order thinking skills. Unfortunately standardized testing does not take cooperative learning into account and does not reward alternative solutions, since there is always only one correct answer on multiple choice tests, thereby penalizing girls.

Tests Biased Against Girls?

There are studies to suggest some bias in standardized testing that favors boys. Ralph Nader (2000), in his article "High-Stakes Tests: Standardized tests produce substandard education" posits that standardized tests "exhibit persistent racial and gender bias, disadvantaging people of color, girls, and women in competitive spheres from primary to graduate school." The basis of this bias is in the way girls think, which is not easily graded by multiple-choice tests.

Sadker and Sadker (1994) have done remarkable work on girls and the inequities they face in American education. In their book *Failing at Fairness; How America's Schools Cheat Girls,* they look at how girls face a gender gap in testing from middle school to medical school. They posit that girls have the early advantage in elementary, but sometime during middle school (echoing Gilligan's work) their grades start to decline, resulting in lower test scores. This in turn blocks them from the most prestigious schools and colleges later on. Females are the only group in American schools that start out ahead and wind up behind.

The standardized test the Sadkers (1994) focused on was the SAT (Scholastic Assessment Test), since that is the only standardized test where an inequity had received publicity as of 1994. On the SAT, boys typically score 50 to 60 points higher than girls. Not only the SAT, but standardized tests at every educational level are biased against girls. But again, in elementary school, girls out-perform boys on every subject, and standardized test. Not until middle school do the boys overtake the girls. The gaps widen as the years progress, with the widest gaps appearing in math and science classes by the end of high school. Grades are not the problem. Boys and girls with identical grades still score differently on the SAT. Boys and girls with A+ averages 83 points lower than a boy with an A+ average. SAT scores that are lower by just a few points can mean not being admitted into a girl's college of choice and can affect the rest of her life.

How exactly are these tests biased against girls? Test bias goes back to World War I, when "mental tests" were given to soldiers in order to match them up to appropriate jobs, but also to prevent certain groups of people from being accepted into elite universities. Needless to say, the test questions were biased in favor of men, the population the test was aimed at. The Army Mental Test included questions such as:

> The Pierce Arrow car is made in
>
> Buffalo
> Detroit
> Toledo
> Flint
>
> Five hundred is played with
>
> Rackets
> Pins
> Cards
> Dice (Sadker & Sadker, 1994, p. 152).

Today's SAT is much improved over the Army Mental Test, but bias remains. The SAT's main purpose is to predict first year success in college, when research shows that GPA (grade point average) is a much better predictor of freshman year success, since grades show stability and commitment to work over time, whereas the SAT is a one-shot deal taken in one morning. However, even with this knowledge, many people still place a higher importance on the SAT than on GPA (Sadker & Sadker, 1994, p. 153). Girls tend to have higher GPAs than SAT scores and not surprisingly, girls have a lower freshman year dropout rate than boys. According to the University of Kentucky Institutional Research Reports Academic Accomplishments of Women (2001), mean ACT scores by gender for the high school graduating class of 2000 in many subjects were higher for boys than girls, including science, math, and composite scores, and girls had a much higher GPA in every class, including overall, yet the high-school dropout rate for boys was 5.76% while the dropout rate for girls was 4.03%. Also, this same study showed that in Britain, the freshman year dropout rate for freshman young men was higher than young women. The British women's GPA were higher than the men's in high school.

This problem is echoed by Sadker and Sadker (1994), who states that while girls are behind boys on scores for standardized achievement tests, their GPAs and report card grades are superior. Is this due to grade inflation on the part of the classroom teachers? Do teachers more frequently tell girls the answers, while requiring boys to figure it out themselves? Are expectations lower for the girls? Are well behaved girls rewarded with grade favors? Sadker and Sadker think so. Research has shown that teachers blend good behavior with grades where girls are concerned (p. 158).

This leads to problems when girls take the SAT, scoring lower than their grades would indicate, and suffering lowered opinions of their intellectual abilities. As a matter of fact, women college students continue to score higher grades than men, although they see themselves as less competent (p. 158).

When schoolteachers lower their expectations for girls, or for any one group of people, it does harm to them by giving them a message that they are unable to perform at the same standards expected of the boys. Since girls internalize the opinions of others, this can create serious problems academically later on.

Good behavior can have it's downside. According to Sadker and Sadker (1994), girls are less likely to be selected for and receive special education services than boys are. The learning difficulties of girls are not identified as often as boys, and as a result, those that could benefit from such services fall farther and farther behind. Behavior problems may be the reason more boys than girls are placed into special ed classes, even though standardized test results indicate only slightly more boys than girls with reading problems, resulting in the best-behaved, quietest girls falling the farthest behind (p. 118).

Head Start

Girls start out their schooling as the favored sex according to the NAEP (National Assessment of Educational Progress), also known as "The Nation's Report Card," that tracks the progress of nine, 13- and 17-year-olds throughout their schooling. This information tells us how girls are doing and how they are promoted. Girls begin school by outperforming boys on almost every single measure. They possess a verbal advantage that carries over into reading and writing, but surprisingly they also hold the advantage in math and social studies. The only area boys hold the advantage is in science (Sadker & Sadker, 1994, p. 138). Do these differences agree with Gilligan's claims that little girls are more vocal and confident that are older girls, with less self-doubt? Or does it reflect teacher's changing notions about girls' abilities as they grow and mature? Is hard work rather than intellectual ability rewarded in girls?

The secret could be the combination of stereotypes teachers hold and to the way standardized tests are constructed. As for the first point, all well-meaning teachers have biases they are not themselves aware of. Let me demonstrate with a simple riddle I use in all of my college classes concerning gender and education:

> A father and son are in a terrible car accident. The father is killed instantly. The son is rushed to the emergency room, when at this point, the surgeon runs into the room, looks at the boy, and says 'I cannot operate on this boy, this boy is my son!' Now how can that be if the father was killed in the car crash?

There is usually silence at this point, as intelligent, enlightened students sit and ponder this puzzle, most without a clue as to what the answer is. Some answers predictably are that the surgeon is the boy's stepfather, or uncle, etc. Only some get it invariably. The answer of course is simple: the surgeon is the boy's *mother.* What makes this answer so invisible to most educated adults? We as a society are so *ingrained* to see roles and expectations in certain ways that our brains become wired and trained to respond accordingly. Even though we are for women's rights, and encourage our female students—the insidiousness of this issue appears to be in our DNA. We are so "in it," we don't "see it," like a fish being in water. Teacher's actions speak much louder than words.

The second issue echoes back to the previously mentioned problem of the cooperative nature of girls' learning styles and the competitive nature of most standardized tests. Girls answer problems in an open-ended, divergent way, with several possible answers, whereas boys tend to answer questions in a more closed-ended convergent way, which is much more suitable for standardized test taking.

CHAPTER 8

WHAT NEXT?

The Future of Science Education in America

This book was born from the frustration I had as a middle school science teacher, parent, college professor, and concerned citizen at the seeming inability of the leaders in our government to understand education at its most basic level. You cannot run a school like a business. Children are not "products," but individuals with varying needs, problems, levels of ability, and gifts. When you assign a "bottom-line" mentality to an educational system, you miss the point of education altogether, which is to open minds to new ideas, and to become a different person than you were when you started your education. (If you are not uncomfortable to a certain degree at some point during your education, you are not getting an education. A *good* education forces you to leave your comfort zone).

Again, I must reiterate here again what I believe, lest I be misunderstood. I *believe* wholeheartedly in high standards for science education and for all subjects. I believe that teachers, students and *parents* should all be held accountable to their part of the deal, or else the two-legged stool topples over. However, from experience and reality checks as a teacher, I know that some teachers are not properly trained in science, and that some parents will not participate in their children's education. If our schools were funded properly, in every section of every city in America, we would have far fewer problems than we do now. A single high-stakes test

The Unfinished Quest: The Plight of Progressive Science Education in the Age of Standards
pp. 107–110

will not address the lacks in these other areas. Nevertheless, the vast majority of those involved in the task of educating our youth care deeply, trying their hardest to make it happen.

Throughout this book, I have stated what I thought to be some of the biggest problems facing education today, namely;

- lack of respect for individual differences in students, including gender, socioeconomic, and racial
- lack of agreement between standards and assessment
- assessing work at the lowest common denominator
- employing standardized tests as high-stakes funding instruments
- assuming passing scores imply learning
- stressing rote learning over creativity and problem solving
- emergence of school as places of high-stress, caused by high-stakes tests
- lack of funding for programs (No Child Left Behind)
- the herding mentality of most school districts
- improper training for teachers in science at the elementary and middle school levels

One of the most basic resources American education needs is proper funding. No Child Left Behind is a noble undertaking, but unfortunately, is underfunded and not assessed properly. A redirection of funds away from foreign concerns and more towards domestic concerns would be a good start. This increased funding would ensure that new assessment instruments with properly trained administrators would once and for all, prove what our children are learning or not learning in our schools. Proper pay for the important work of teaching, the most underrespected profession in America, would also ensure properly trained teachers, in all subjects and disciplines.

The most important aspect of any education, learning how to think, is falling out of fashion. We overlook creativity, the highest level of learning, at our peril. Nothing is more important in solving real world problems as they arise, whether in industry, science, medicine, business, or any other field. Mihaly Csikszentmihalyi (1996) has written much on the subject of creativity, including *Creativity: Flow and the Psychology of Discovery and Invention*. One of the points he stresses is that

> most of the things that are interesting, important, *human* are the results of creativity. We share 98% of our genetic makeup with chimpanzees. What makes us different—our language, values, artistic expression, scientific understanding, and technology—is the result of individual ingenuity that

> was recognized, rewarded, and transmitted through learning. Without creativity, it would be difficult indeed to distinguish humans from apes. (PAGE NUMBER(S)?)

Real *creative* solutions to what ails education are not cheap and easy, as are bubble tests. Much of America's effort towards a globally competitive school system is admirable, but it becomes misguided when it attempts to treat students as products or bottom lines, so numbers can be crunched for comparative purposes.

In this product (grade) driven society in which we live, one has only to remember what Dewey (1916) said about learning; about process being more important than product:

> To keep the process (of learning) alive, to keep it alive in ways which make it easier to keep it alive in the future, is the function of educational subject matter ... an individual can only live in the present. The present is not just something which comes after the past; much less something produced by it. It is what life is in leaving the past behind it. The study of past *products* will not help us understand the present. (p. 75)

Progressive education may have its limitations, as do all philosophies. According to Gardner (1991), one of the main problems has been a neglect of assessment aimed towards Progressive education. Gardner states,

> In most traditional forms of education, assessments are common-indeed, all too common ... they are often inappropriate and simplistic, tending to lead away from deeper forms of understanding. Thus it was entirely understandable-indeed praiseworthy-that many progressive educators spurned formal instruments, and where necessary, sought to obtain waivers from the tyranny of standardized tests. (p. 197)

However, even with this challenge, Gardner also has faith in progressive education. "Indeed, it is in the most fully articulated models of progressive education that I find clues toward the construction of an educational environment in which genuine understandings can become a reality" (p. 199).

There are countless students in American schools today, from kindergarten to graduate school, who care more about their grades than they do about whether they learned anything valuable or not. As a result, teachers and professors become a student's "employee" instead of mentor and teacher. Rampant grade inflation has occurred because college students demand As, and an untenured professor's job hangs in the balance as part of their promotion decision depends on evaluations from students.

This is our fault—this pressure comes from us, their so-called leaders, who stress passing tests and grades over true knowledge and make education a competitive endeavor rather than a striving of the individual to better one's own mind for the contribution towards society and the betterment of humankind. One solution would be a three-part National Standardized Test, mentioned earlier in this book, which would address national, state, and local issues, and would better differentiate for cultural differences across the country. Also, proper funding for these assessments and less stress on the high-stakes nature of the tests themselves. Education is about the process, not the product. We must forgo this notion of churning out students who care only about what grade they made or what their score was on a standardized tests, but instead lead the way and focus on training children how to think, how to criticize, how to deduct, how to problem solve, how to figure, how to argue, how to create, how to *appreciate* —only then will our educational "product" be superior to that of any other nation on earth, for one cannot have leadership without leaders.

REFERENCES

A Nation at Risk, The Imperative for Educational Reform. (1983). United States Government, 8-11.

Adams G. L., & Englemann, S. (1996). *Research on direct instruction: 25 years beyond DISTAR*. Seattle, WA: Educational achievement Systems.

Addam, J. ((1945). Twenty years at Hull House. New York: MacMillan.

American Association of Colleges for Teacher Education. (2005). Retrieved 2005, from http://www.aacte.org

American Youth Policy Forum. (2000, January 7). *The impact of high stakes testing policies on minority and disadvantaged students: A forum.* Retrieved 2005, from www.aypf.org/forumbriefs/2000/fb010700.htm

Andrews, J. F. (1985, June). *Deaf children's acquisition of prereading skills using the reciprocal teaching procedure.* Paper presented at the Council of American Instructors of the Deaf, Florida School for the Deaf, St. Augustine, Florida.

Applefield, J. M., Huber, R., & Moallen, M. (2000-2001, December-January). Constructivism in theory and practice: Toward a better understanding. *The High School Journal, 84*(2), 34-53.

Armstrong, T. (1994). *Multiple intelligences in the classroom.* Alexandria, VA: Association for Supervision and Curriculum Development.

Armstrong, T. (1998). *Awankening genius in the classroom.* Alexandria, VA: ASCD.

Appropriate use of high-stakes tests in our nation's schools. (2001). *APA online.* Retrieved 2005, from http://www.apa.org/pubinfo/testing.html

Ausubel, D. (1963). *The psychology of meaningful verbal learning.* New York: Greene & Stratton.

Asante, M. K. (1998). *Afrocentricity.* Trenton, NJ: Africa World Press.

Bailey, S. M. (Ed.). (1992). *How schools shortchange girls: The AAUW Report, a study of major findings on girls and education*. New York: Marlowe.

Bandura, A. (1977). *Social learning theory.* Upper Saddle River, NJ: Prentice Hall.

Barger, R. N. (2004). *Impact Of business and industry.* Retrieved from http://www.nd.edu/~rbarger/www7/impbusin.html

Barman, C. R. (1989). A procedure for helping prospective elementary teachers integrate the learning cycle into science textbooks. *Journal of Science Teacher Education, 1*(2), 21-26.

Bay, M., & Others. (1992). Science instruction for the mildly handicapped: Direct instruction versus discovery teaching. *Journal of Research in Science Teaching, 29*(6), 555-570.

Berger, S. L. (1999). *Opening the gate: Changing the attitudes and practices of teachers through a constructivist professional development model* Unpublished doctoral dissertation, Florida State University.

Berube, C. T. (2000). A conceptual model for middle school science instruction. *The Clearing House, 73*(6), 312.

Berube, C. T. (2004) Are standards preventing good teaching? *The Clearing House,* 77(6), 264-267.

Berube, C. T. (2007). Autism and the artistic imagination: The link between visual thinking and intelligence. *Teaching Exceptional Children Plus, 3*(5). Retrieved from http://escholarship.bc.edu/education/tecplus/vol3/iss5/art1/

Berube, M. (2005, December 5). Virginia should drop SOL testing. *The Virginian Pilot.*

Berube, M. R. (1994). *American school reform: Progressive, equity, and excellence movements, 1883-1993.* Westport, CT: Praeger.

Bevevino, M. M., Dengel, J., & Adams, K. (1999). Constructivist theory in the classroom: Internalizing concepts through inquiry learning. *The Clearing House, 72(*5), 274-275.

Bloom, B. S. (Ed.). (1956). *Taxonomy of educational objectives, the classification of educational goals, handbook I: Cognitive domain.* New York: David McKay.

Bol, L., & Strage, A. A. (1996). High school biology: What makes it a challenge for teachers? *Journal of Research in Science Teaching, 33*(7), 125, 753-772.

Bol, L., Stephenson, P. L., & O'Connell, A. A. (1998). Influence of experience, grade level, and subject area on teachers' assessment practices. *The Journal of Educational Research, 91*(6), 323-330.

Bowers, R. S. (1991). Effective models for middle school science instruction. *Middle School Journal, 22*(4), 4-9.

Brooks, J. G., & Brooks, M. G., (1993). *The case for constructivist classrooms.* Alexandria, VA: Association for Supervision & Curriculum Deve.

Burris, C. (2003). Creating equitable high schools: Strategies to eliminate tracking and ability grouping. *American Youth Policy Forum.* Retrieved December 12, 2003, from http://www.aypf.org/forumbriefs/
2003/fb121203.htm

Brewmen, S. (2003). The brain game. *The Daily Star.* Retrieved September 22, 2005, from http://www.thedailystar.net/magazine/2003/11/01/health.htm

Bruner, J. (1960). *The process of education.* Cambridge, MA: Harvard University Press.

California Educator. (2004). Retrieved April 14, 2008, from http://www.cta.org/media/publications/educator/archives/2004/200409_feat_01.htm

Chang, M. M. (1994). *Constructivist and objectivist approaches to teaching chemistry concepts to junior high school students.* Paper presented at AERA, New Orleans. LA.

Chen, Z. (1999). Schema induction in children's analogical problem solving. *Journal of Educational Psychology, 4,* 703-715.

Clough, M. P. (2000). The nature of science: understanding how the game of science is played. *The Clearing House, 74*(1), 14.

Associated Press. (2005). Exit exams deny diplomas. *CNN.com Education.* Retrieved from http://www.cnn.com/2003/EDUCATION/06/15/exit.exams.ap/index.html

Csikszentmihalyi, M. (1996). *Creativity; flow and the psychology of discovery and invention.* New York: HarperCollins.

Daas, P. M. (2000). Preparing coaches for the changing game of science: teaching in multiple domains. *The Clearing House, 74*(1), 39-41.

Darder, A. (1991). *Culture and power in the classroom: A critical foundation for bicultural education.* Westport, CT: Bergin & Garvey.

de Esteban, M., & Penrod, K. (2000). *Effect of teaching philosophy orientation on students' levels of communication apprehension* (Research report No. 143). Brookings, SD: South Dakota St. University, Department of Undergraduate Teacher Educaton, College of Education and Counseling.

Dewey, J. (1916). *Democracy and education.* New York: The Free Press.

Diamond, M., & Scheibel, A. B. (1985). *The human brain coloring book.* New York: HarperCollins.

Discenna, J., & Howse, M. (1998). *Biology and physics students' beliefs about science and science learing in non-traditional classrooms.* Paper presented at AERA, San Diego, CA.

Driver, R. (1989). The construction of scientific knowledge. In R. Miller (Ed.), *Doing Science: Images of science in science education* (pp. 83-106). London: Falmer Press.

Dunkhase, J. A., Hand, B. M., Shymansky, J. A., & Yore, L. D. (1997). *The effect of a teacher enhanced project designed to promote interactive-constructivist teaching strategies in elementary school science on students' perceptions and attitudes.* Paper presented at the School of Science and Mathematics conference, Milwaukee, WI.

Ebenezer, J. V., & Haggerty, S. M. (1999). *Becoming a secondary school science teacher.* Upper Saddle River, NJ: Prentice Hall.

Education law gets first test in U.S. schools. (2005, October 20). *The New York Times,* p. A19.

Egan, K. (2002). *Getting it wrong from the beginning: Our progressive inheritance from Herbert Spencer, John Dewey, and Jean Piaget.* London: Yale University Press.

Encyclopedia Britannica Online. (1999). Retrieved September 23, 2005, from http://search.eb.com/women/article-9124974

Encyclopedia Laborlawtalk.com. (2005). Retrieved September 23, 2005, from http://encyclopedia.laborlawtalk.com/dysgraphia

Finn, C., & Ravitch, D. (1996). *Education Reform 1995-1996: A report from the educational excellent network to its education policy committee and the American people.* Retrieved from www.edexcellence.net/library/epctot.html

Fisher, H. (n.d.). *Anthropologist.* New Jersey: Rutgers University. Retrieved www.helenfisher.com

Fosnot, C. T. (1989). *Inquiring teachers, inquiring learners: A constructivist approach for teaching.* New York: Teachers College Press.

Gallagher, J. J. (1991). Prospective and practicing secondary school science teachers' knowledge and beliefs about the philosophy of science. *Science Education, 75,* 121-133.

Garcia, F. (2004). Developing sociopolitical literacy: intellectual consciousness for urban middle schools communities. *The Clearing House, 34*(7), 34-40.

Gardner, H. (1983). *Frames of mind: the theory of multiple intelligences.* New York: HarperCollins.

Gardner, H. (1991). *The unschooled mind: How children think and how schools should teach.* New York: Harper Collins.

Gega, P. C., & Peters, J. M. (1998). *Science in elementary education.* Upper Saddle River, NJ: Merrill.

Gilligan, C. (1993). *In a different voice: Psychological theory and women's development.* Cambridge, MA: Harvard University Press.

Grandin, T. (2003, Spring). Genuis may be an abnormality: Educating students with Asperger's Syndrome, or high functioning autism. *Paradigm.*

Grandin, T. (n.d.). *Dr. Temple Grandin's Web Page.* Retrieved May 10, 2006, from http://www.grandin.com/

Greene, B. (2000). *The elegant universe: Superstrings, hidden dimensions, and the quest for the ultimate theory.* New York: Vintage Books.

Guyver, R. (1999). *National curriculum: Key concepts and curriculum controversy.* Retrieved from www.Harris8.freeserve.co.uk/rguyver.html

Hambleton, R., Jaeger, R. M., Koretz, D., Linn, R. L., Millman, J., & Phillips, S. E. (1995). *Review of the measurement quality of the Kentucky instructional results information system 1991-1994* (Report prepared for the Kentucky General Assembly.) Frankfort, KY: Office of Educational Accountability.

Hanley, S. (1994). *On constructivism.* College Park, MD: Maryland Collaborative for Teacher Preparation.

Heide, C. L. (1998). *Attitudes of eighth grade honors students toward the conceptual change methods of teaching science (middle school students.* Unpublished doctoral dissertation, Northern Arizona University.

Howard, B. C., McGee, S., Schwartz, & N., Purcell S. (2000). The experience of constructivism: transforming teacher epistemology. *Journal of Research on Computing in Education, 32*(4), 455.

Howe, A. C., & Jones, L. (1993). *Engaging children in science* (1st ed.). New York: MacMillan.

Hoyt, A. (Producer, Writer, & Director) (1999). *The richest man in the world andrew carnegie* [Film]. United States: PBS American Experience.

Hyde, J. S. (1996). *Half the human experience: The psychology of women* (5th ed.) Washington, DC: Heath & Co.

Individuals With Disabilities Education Act. (2005). *Council for exceptional children: The voice and vision of special education.* Retrieved from http://www.cec.sped.org/law_res/doclaw/index.php

Institutional Research Reports. (2001, September). *University of Kentucky, Academic Accomplishments of Women*. Retrieved 2005 from http://www.uky.edu/IR/dsi/indicators/women.pdf

Isaacson, W. (2007). *Einstein: His life and universe*. New York: Simon & Schuster.

Jeter, A. (2005, September 21). Norfolk wins top award for urban education. *The Virginian-Pilot,* pp. A1, A14.

Jeter, A. (2007, June 3). Why Are So Many Students Acing Some SOL Tests? *The Virginian-Pilot,* pp. A1, A10, A11.

Johnson D. W., & Johnson, R. T. (1994). *Learning together and alone: Cooperative, competitive, and individualistic learning* (4th ed.) Boston: Allyn & Bacon.

Kaplan Test Prep. (2005). *Web site.* Retrieved from http://www.kaptest.com/

Kozloff, M. A., LaNunziata, L., & Cowardin, J. (1999), *Direct instruction in education*. Retrieved 2001 from www.uncwil.edu/people/Kozloffm/diarticle.html

Kozloff, M. A., LaNunziata, L., Cowardin, J., & Bessellieu, F. B. (2001). Direct instruction: Its contribution to high school achievement. *The High School Journal, 84*(2), 54-70.

Kozol, J. (2005). *The shame of the nation: The restoration of apartheid schooling in America.* New York: Crown Publishers, Random House.

Lave, J., & Wenger, E. (1991). Authors, text and talk: The internalization of dialogue from social interaction during writing. *Reading Research Quarterly, 29,* 201-231.

Lewis, A. (2003). Beyond testing. *The Education Digest, 68*(1), 70-71.

Marchland-Martella, N. E., Martell, R. C., & Lingnugaria-Kraft, B. (1997). Observation of direct instruction teaching behavior. Determining a representative instruction teaching for supervision. *International Journal of Special Education, 12*(2), 32.

McCombs, J. S., & Carroll, S. J. (2005). Ultimate test: Who is accountable for eduction if everybody fails? *Rand Review, 29*(1), 10.

McCombs, J. S., Kirby, S. N., Barney, H., Darilek, H., & Magee S. (2004). *Achieving State and national literacy goals, a long uphill road: A Report to Carnegie Corporation of New York*. Retrieved April 14, 2008, from Rand Corporation Web site: http://rand.org/pubs/technical_reports/TR180-1/

McDermott, L. C. (1993). How we teach and how students learn. *Annals of the New York Academy of Science, 701,* 9-19, 295.

McNichols, T. J. (2000). Deconstructing constructivism: The Kantian connection. *The Journal of Philosophy and History of Education, 49,* 4.

Meade, A. (1999, November). *Schema learning and its possible links to brain development.* Paper presented at at a seminar at the Children's Hospital of Michigan, Wayne State University.

Members of the First Cohort of the BSU doctoral program for TE660. www.coehp.idbsu.edu/FACHTMLS/cohort3/ausebel.htm

Miriam-Webster's. (n.d.) *Online Dictionary.* http://www.m-w.com/cgi-bin/dictionary?book=Dictionary&va=progressive&x=13&y=13

Moore, I. (2003, Feb.). Does the state standardized testing program help or hurt low income students and students of color? *Action Alliance for Children:*

Children's Advocate. Retrieved from http://www.4children.org/news/103imo.htm

Morrison (2003). Retrieved from http://infoeagle.bc.edu/bc_org/avp/law/lwsch/journals/bclawr/41_5/03_TXT.htm

Musheno, B. V., & Lawson, A. E. (1999). Effectsd of learning cycle and traditional texts on comprehension of science concepts by students at differeing reasoning levels. *Journal of Research in Science Teaching, 36*(1), 23-37.

Nader, R. (2000, November 6). High-Stakes Tests: Standardized tests produce substandard education. *The San Fransico Bay Guardian.*

National Conference for State Legislators. (2005). *State legislators offer formula for improving No Child Left Behind Act.* Retrieved from http://www.ncsl.org/programs/press/2005/pr050223.htm

National Research Council. (2005). *Fairtest: The national center for fair and open testing.* Retrieved from http://www.fairtest.org/examarts/fall98/k-nrc.htm

Newbold, C. T., (1999). Multiple intelligences and the artistic imagination: A case study of Einstein and Picasso. *The Clearing House*, *72*(3), 153-155.

Noll, J. Wm. (2005). *Taking sides: Clashing views on controversial educational issues* (13th ed.). Columbus, OH: McGraw-Hill/Dushkin.

O'Neill, P. T., J. D., Willkie, Farr & Gallagher. (2000). *Is testing legal? 5 Key factors to look for when high-stakes graduation tests are imposed on kids with LD.* Retrieved from http://www.ldonline.org/ld_indepth/assessment/oneill_fivepoints.html

Orfield, G. (2005). *The Civil Rights Project.* Harvard University. Retrieved April 11, 2008, from http://www.civilrightsproject.harvard.edu/resources/civilrights_brief/testing.php

Palinscar, A., David, Y., & Brown, A. (1984). *Reciprocal teaching of comprehension-fostering and comprehension-monitoring activities. Cognition and Instruction, 1*(2), 121-122, 145, 153, 164.

Palinscar, A., David, Y., & Brown, A. (1992). Using reciprocal teaching in the classroom: A guide for teachers. *The Brown/Campione Research Group, 12,* 42-47.

Parson, J., & Polson, D. (1931). *Siegfried Engelmann.* Retrieved April 11, 2008, http://psych.athabascau.ca/?html/387/Openmodules/Engelmann/Engelmannbio.shtml

Pollard, J. (2005). Poor teaching for poor kids (Review). *Standardized Testing.net.* Retrieved from http://www.standardizedtesting.net/teaching.htm

Ravitch, D. (1996). The case for national standards and assessments. *The Clearing House, 69*(3), 134-135.

Ravitch, D. (2005, November 7). Every state left behind. *New York Times,* p. A25.

Richardson, V. (1999). Teacher education and the construction of meaning. In G. Griffin (Ed.), *The education of teacgers: Ninety-eight yearbook of the National Society for the Study of Education* (Part 1, pp. 46, 145-166). Chicago: University of Chicago Press.

Riis, J. (1890). *How the other half lives.* New York: Charles Scribner's Sons.

Rutter, M. (2005). Incidence of autism spectrum disorders: Changes over time and their meaning. *Acta Paediatr, 1,* 2-15.

Sadker, D., & Sadker, M. (1994). *Failing at fairness; How America's schools cheat girls.* New York: Charles Scribner's Sons.

Schroeder, K. (2003). High-stakes horrors. *The Education Digest, 68*(9), 54-55.

Shepard, L. A. (2000). The role of assessment in a learning culture. *Educational Researcher, 29*(7), 4-14.

Slavin, R. (1985). An introduction to cooperative learning research. In *Learning to cooperate, cooperating to learn*. New York: Plenum Press.

Sobel, T. (1991, July 12). Understanding diversity. *Memorandum to New York State Board of Regents,* p. 7.

Sowell. T. (1997). *Late talking children.* New York: Basic Books.

Sowell. T. (2001). *The Einstein syndrome: Bright children who talk late.* New York. Basic Books.

Steele, C. M. (1999). Thin ice: Stereotype threat and black college students. *The Atlantic Monthly, 284*(2), 44-7, 50-4.

Strage, A. A., & Bol, L. (1996). High school biology: What makes it a challenge for teachers? *Journal of Research in Science teaching, 33*(7), 753-772.

Taba, H. (1932). *The dynamics of education: A methodology of progressive educational thought. London:* Kegan Paul, Trench, Trubner & Co.

Taba, H. (1966). *Teaching stragegies for the culturally disadvantaged.* Chicago: Rand McNally.

Taylor, F. W. (1911). *The principles of scientific management.* New York: Harper & Brothers.

Taylor, P. C., Fraser, B. H., & White, L. R. (1994). *CLES: An instrument for monitoring the development of constructivist learning environment.* Paper presented at AERA, New Orleans, LA.

Thomasini, N. G., & Others. (1990). *Teaching strategies and conceptual change: sinking and floating at elemtary level.* Paper presented at AERA, Boston.

Tyler, A. (1999). Does tracking resegregate schools? *Rural Policy Matters, 1*(5), Retrieved from http://www.ruraledu.org/rpm/rpm105b.htm

United States Department of Education. (2005). *Goals 2000: Educate America Act.* Retrieved from http://www.ed.fov/legislation/GOALS2000/TheAct/sec/02.html

Urban, W. J., & Wagoner, J., Jr. (1996). *American education: A history.* Columbus, OH. McGraw-Hill.

Virginia: Standards of Learning Assessments. (n.d.). *Statewide Spring passing rates.* Retrieved February 1, 2006, from http://www.pen.k12.va.us/VDOE/Assessment/StatePassRates02.html

Virginia: Standards of Learning Assessments. (n.d.). *Statewide Spring passing rates.* Retrieved February 1, 2006, from http://www.pen.k12.va.us/VDOE/Superintendent/Sols/science7.doc

Walker, C. (1999). *The effect of different pedagogical approaches on mathematics students achievement.* Paper presented at AERA, Montreal, Canada.

Welner, K. (2003). Creating equitable high schools: Strategies to eliminate tracking and ability grouping. *American Youth Policy Forum.* Retrieved December 12, 2003, from http://www.aypf.org/forumbriefs/2003/fb121203.htm

Wikipedia.com. (n.d.). Retrieved from http://en.wikipedia.org/wiki/Learning_disabled

Wilson, W. J. (1997). *When work disappears: The world of the new urban poor.* New York: Vintage Books

Yarlas, A. S. (1999). *Schema modification and enhancement as predictors of interest; at test of the knowledge-schema theory of cognitive interest.* Paper presented at AERA, Montreal, 29-35.

Yager, R. E. (Ed.) (1997). *SALISH: A research project dedicated to improving science and mathematics teacher education. Secondary science and mathematics teacher preparation programs: Influences on new teachers and their students.* Washington, DC: U.S. Department of Education.

Printed in the United States
114835LV00002B/3/P